Subtraction

4TH & ABOVE

Designs in Math

WRITTEN & ILLUSTRATED BY

RANDY L. WOMACK, M.ED.

LEARNING DISABILITIES & ART

PUBLISHED BY

G.E.C. PUBLICATIONS golden educational CENTER

"LEADING THE WAY IN CREATIVE EDUCATIONAL MATERIALS" ™

857 LAKE BLVD. ❖ REDDING, CALIFORNIA 96003

Teacher – Parent Instructions For *Designs in Math*

This book has been written to aid the classroom teacher or parent in activities that reinforce math facts while giving the student practice drawing with a straight edge. *Designs in Math* can be introduced any time after a student fully understands the particular math function in the book.

It is extremely important that the teacher does not take for granted that students know how to properly draw lines without frustration using a straight edge or ruler. Many children have a great deal of difficulty with this skill because they are not properly instructed or do not have ample opportunity to practice. If one of your students has a lot of difficulty drawing with a straight edge, it might be to their benefit to begin with the easier books — *Creating Line Designs.* (See the back of this book for information regarding this series, as well as *Beginning Math Art.*)

The directions on each page have been simplified in a step-by-step format so that students can easily read and understand them independent of teacher instruction. However, it is recommended that the teacher spend some initial time on the directions and straight edge techniques. "Tips on Using a Straight Edge" have been included in this book to aid teachers in presenting the use of a straight edge to students. After this initial time is spent, children will be able to complete all of the designs in this book with very little teacher instruction.

These Books Develop

- Math Fact Memorization
- Visual Discrimination
- Following Directions Skills
- Eye–Hand Coordination
- Fine Muscle Control
- Basic Design Principles
- Logic
- Sequential Thinking
- Right Brain Functioning

Published By Golden Educational Center
857 Lake Blvd. ❖ Redding, California 96003

ISBN 1–56500–005–6

Tips on Using a Straight Edge

In teaching children how to use a ruler or straight edge, it is important that the following points be included:

1. The child should hold his hand out with the palm down on the table.

2. He should hold the straight edge with the tips of his thumb and fingers — as far apart as is comfortable. (Note: Most children will hold it in the middle, and the ends will move as they draw their lines.)

3. The child should hold his pencil at a perpendicular angle. (Note: Many children will place the pencil point at a 45–30° angle into the edge of the ruler and usually lift the ruler up with the point of the pencil.)

4. Be sure the child places his pencil on one of the dots before he moves the straight edge to that dot. After he has his pencil on one of the dots, he can slide the other end of the straight edge to the dot he wants to connect. This is an essential step to be learned by the student. It is the most accurate way to line up the straight edge with the two dots needing to be connected — and accuracy is an important element in the child feeling successful about his design.

5. Always have the child move his pencil from top to bottom and from the left to the right. (Left handed children would move from top to bottom and then right to left.) This will add to the accuracy of the lines. The lines will also be easier to make and neater in appearance.

6. Instruct the child to press gently on the pencil. (Note: Most children will want to press down hard, which causes the straight edge to move, as well as hand fatigue.)

7. Some children may reach over the straight edge to connect the dots, using the awkward edge. Instruct them to draw on the right and top edge of the ruler if they are right handed. (A left-handed child should only draw on the left and top edge of the ruler.)

8. To avoid reaching over the ruler, the student may turn his paper, instead of the straight edge, to a comfortable orientation to facilitate his drawing the designs.

9. Some children (usually the younger ones) may find it easier to stand up while working on their designs.

10. The child must always complete connecting the dots in numerical order then alphabetical order. If they do not, the designs will not work and they will get frustrated to no end — but that is part of the learning process.

✎ Please Note: We wrote the above information referring to a male student. We like girls, too. However, we thought it easier reading to use one gender rather than he/she.

Some students may want to do one or more of the following activities.

Suggestions for Additional Activities

1. Children enjoy using felt pens for coloring the designs. However, every so often, challenge the students to experience different media such as colored pencils, chalk, "artista" oils and even water color paints.
2. Most children will want to use many colors or just certain favorites. Encourage each student to experiment with color patterns of only two or three colors — a great deal of color wheel reinforcement is possible.
3. After he colors the completed design, have the child re-draw the lines with a black pen (when the media permits).
4. Have the child cut out completed designs and paste them on colored background paper. He could even frame the finished product(s).
5. After he colors a design, have the child cut out the specific shapes and paste them onto a colored piece of paper in the exact design seen on the respective "correction key." Many of the designs are difficult. Be certain that the child is capable of doing the chosen design so he doesn't get frustrated.
6. After he colors a design, have the child cut out the specific shapes and paste them onto a colored piece of paper to create a unique design of his own.
7. **Introduce students to string art:**
   a. Several of the designs in this book lend themselves to the craft of string art.
   b. Direct the student to choose one of his favorite designs. (Use your judgement as to the difficulty factor involved.) Have the child complete the design on paper. He does not need to color it.
   c. Give the child a square piece of three-quarter inch thick soft wood (at least 8" square). Plywood is inexpensive and easy to use. It also looks nice when painted or stained a color.
   d. Tape his completed design to the wood with masking tape.
   e. Choose the appropriate nails to work with. (1"–1½" long work well)
   f. Now, direct the student to hammer nails so that about half of each nail remains above the board. Make certain that the nails are placed directly over the dots on the paper taped to the wood.
   g. Have the student tie and run string from one nail to the other until the design is completed. (The stringing may have to be altered from the exact design in order to work. However, encourage the child to allow for design alterations in order to make the stringing easier.)
   h. After the stringing is completed, carefully remove the paper taped to the wood and you have a fine example of string art.

8. **Introduce students to nail sculpture:**

   a. Several of the designs in this book lend themselves to the craft of nail sculpture.

   b. Direct the student to choose one of his favorite designs. (Use your judgement as to the difficulty factor involved.) Have the child complete the design on paper. He does not need to color it.

   c. Give the child a square piece of soft wood (at least 8" square). It must be at least three-fourths inch thick. However, one and one-half inch thick works very well. Try using a rough-sawn exterior wood such as cedar. (Plywood does not work well for this project unless it is hung indoors.) Staining or painting the wood before it is used is optional.

   d. Tape his completed design to the wood with masking tape.

   e. Choose a variety of different nails and screws to use. Make sure to choose some that will rust and some that will not rust (galvanized do not rust.) Get a variety of lengths and head sizes. Brass screws are expensive, but they add a special dimension to the sculpture when it is completed.

   f. Now, direct the student to hammer the nails (or screws), filling an entire shape within the design. (Screws can be hammered very easily. However, screwing them in is excellent for muscle development. He may need instructions on how to make a starter hole with a nail or "Yankee" screwdriver.) Explain that it is important to put the nails as close together as possible in order to create the design effectively. (Some of the shapes within the design may be omitted or changed as needed.)

   g. It is important for the child to use the same type of nail for each of the specific shapes chosen. However, it is equally important for him to vary the different nails within the whole design. The student should also be instructed to vary the depth he hammers the nails into the wood from shape-to-shape. Graduated depths within the same shape create a very nice result.

   h. After the child has filled the desired shapes (some may be left with just the wood showing), get as much of the paper out as possible and have the child hang his finished sculpture outside to be rusted and changed by the weather. The paper will eventually deteriorate.

9. Some students may want to create their designs and have other students complete them.

Correction Key
Design #1

1. Use this design as a correction key.
2. Allow your students to correct their own work.
3. Make a transparency of this design and instruct your students to place the transparency over their completed design for a quick and easy check.

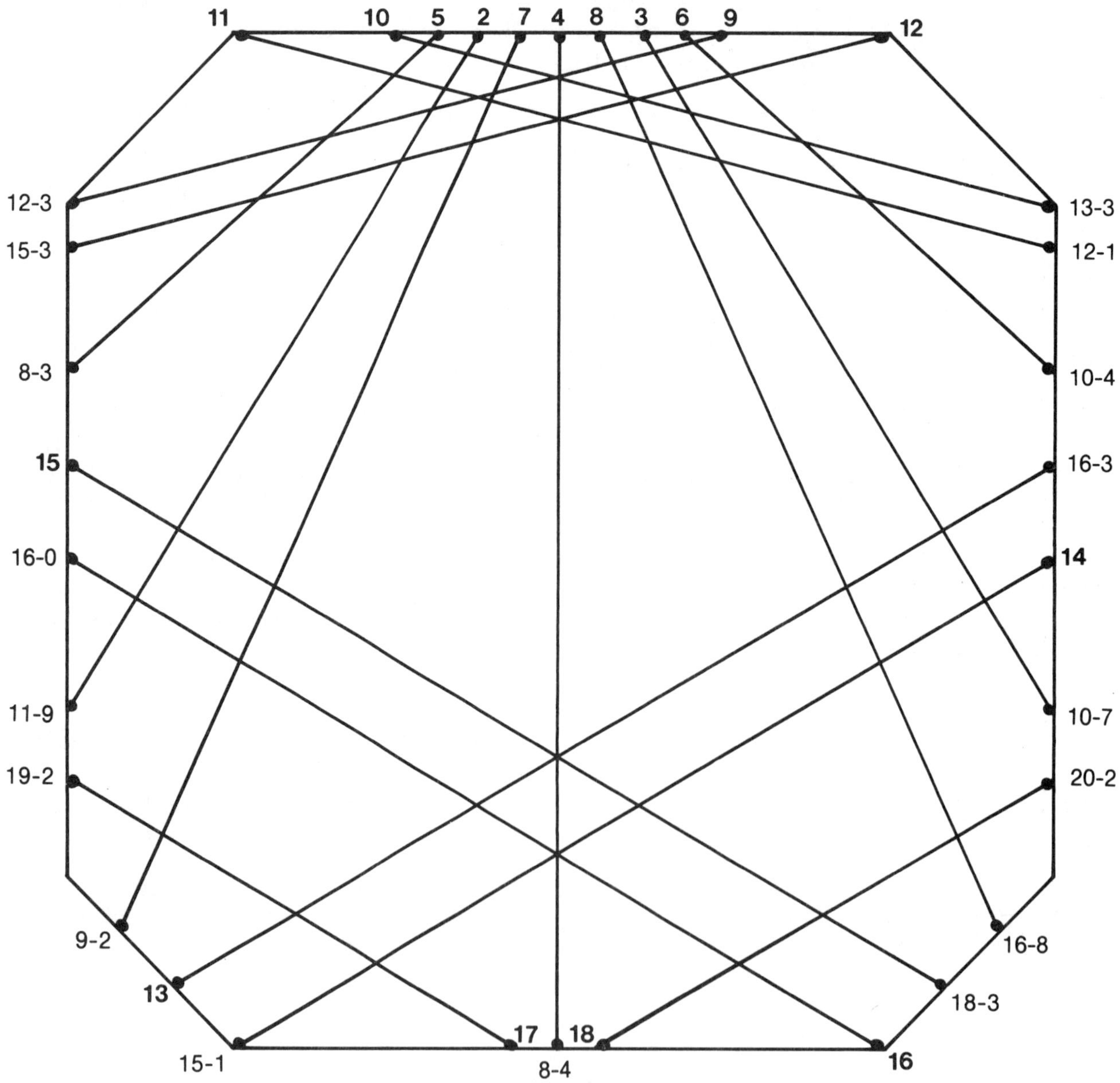

Subtraction
Design #1

Name ______________________________

Date ______________________________

1. **Read all of the directions before you begin.**
2. **Complete one subtraction problem next to a dot.**
   (You may use another piece of paper if you need to.)
3. **After you complete the problem, find the answer on the paper.**
4. **Use a ruler (or straight edge) and draw a line to connect the dot by the problem to the dot by the correct answer.**
5. **Complete each of the problems in the same way.**
6. **Carefully color your design after completing *all* of the problems.**

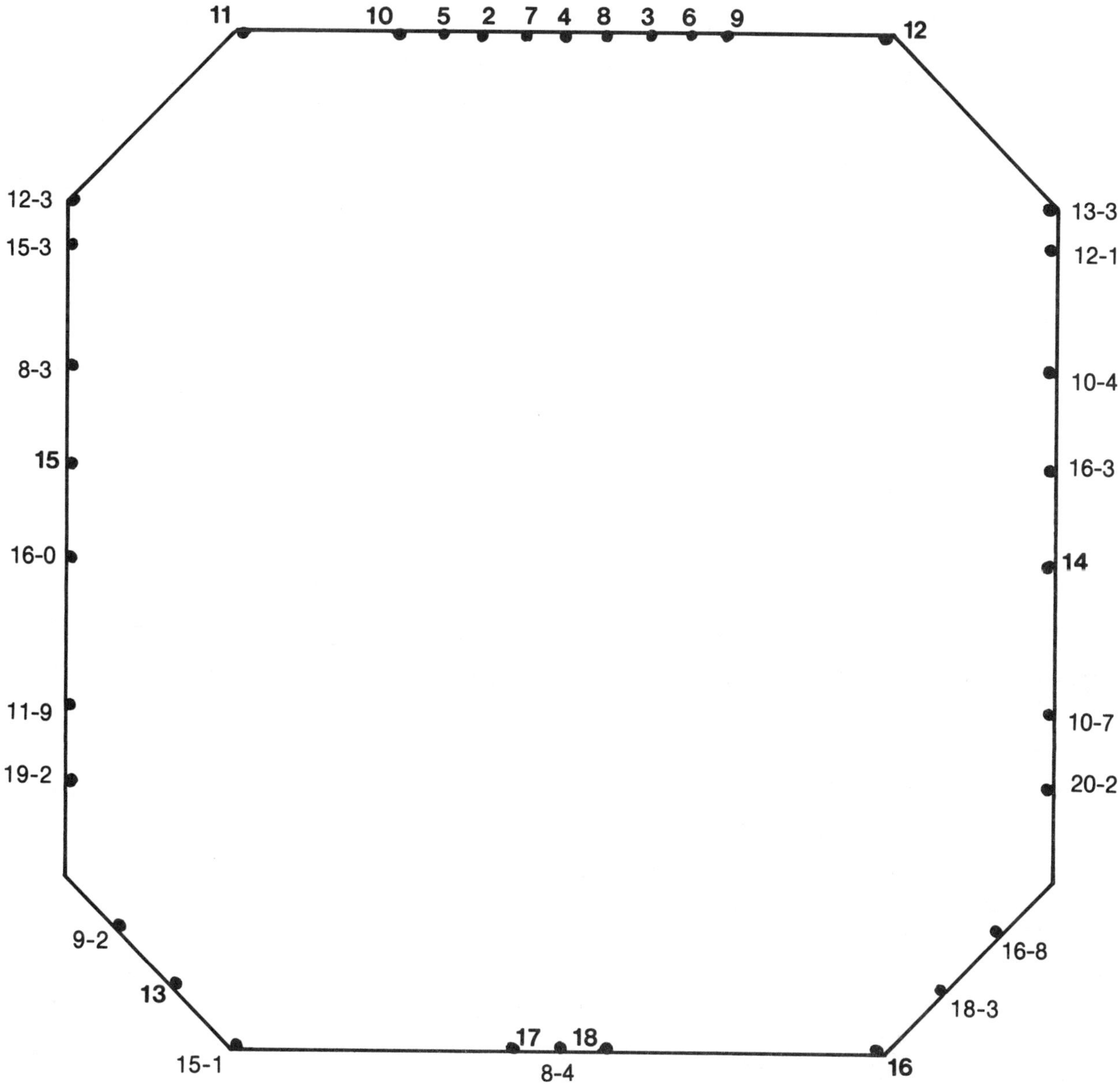

Correction Key
Design #2

1. Use this design as a correction key.
2. Allow your students to correct their own work.
3. Make a transparency of this design and instruct your students to place the transparency over their completed design for a quick and easy check.

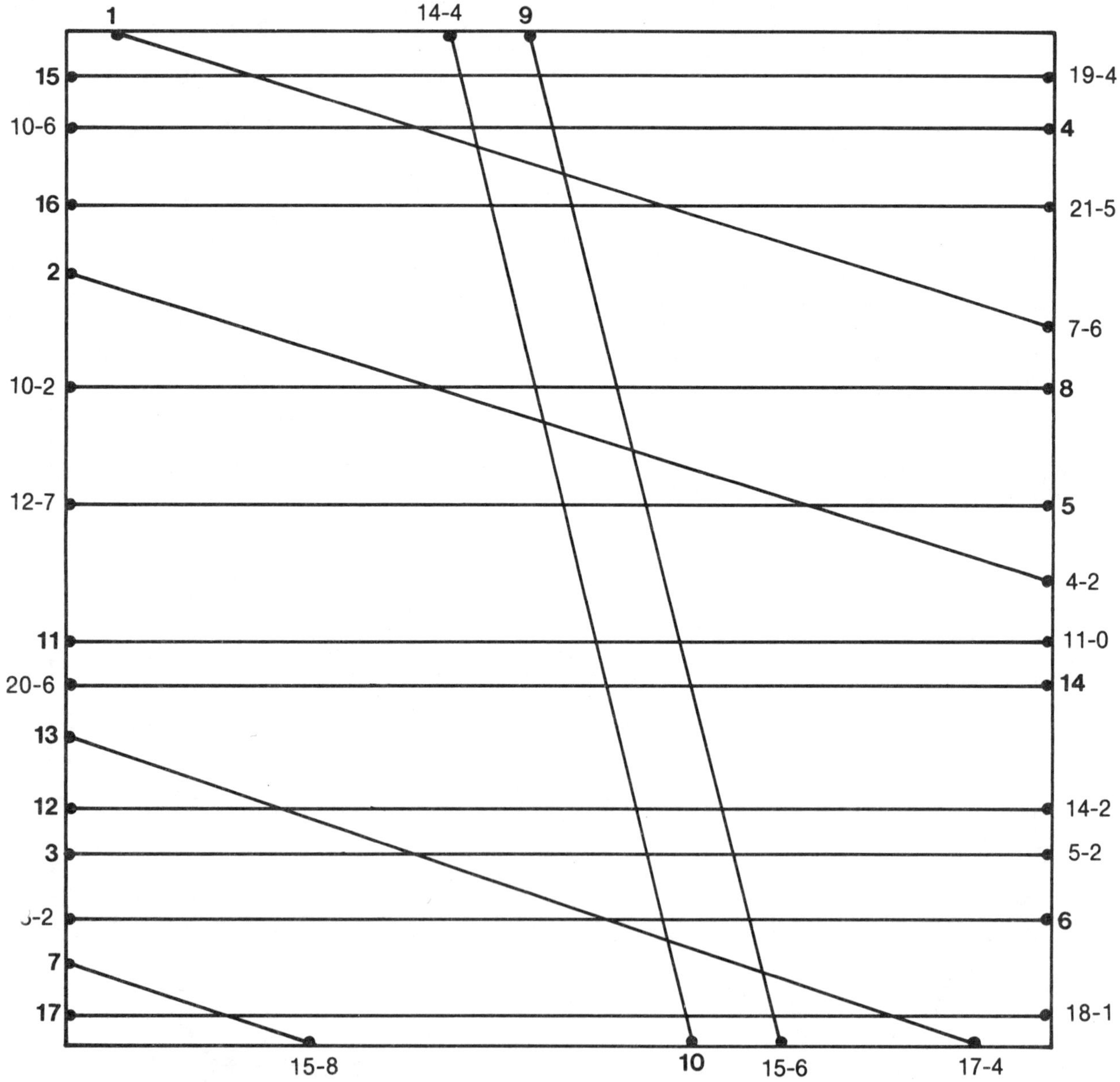

Subtraction

Name ______________________

Design #2

Date ______________________

1. **Read all of the directions before you begin.**
2. **Complete one subtraction problem next to a dot.**
   (You may use another piece of paper if you need to.)
3. **After you complete the problem, find the answer on the paper.**
4. **Use a ruler (or straight edge) and draw a line to connect the dot by the problem to the dot by the correct answer.**
5. **Complete each of the problems in the same way.**
6. **Carefully color your design after completing *all* of the problems.**

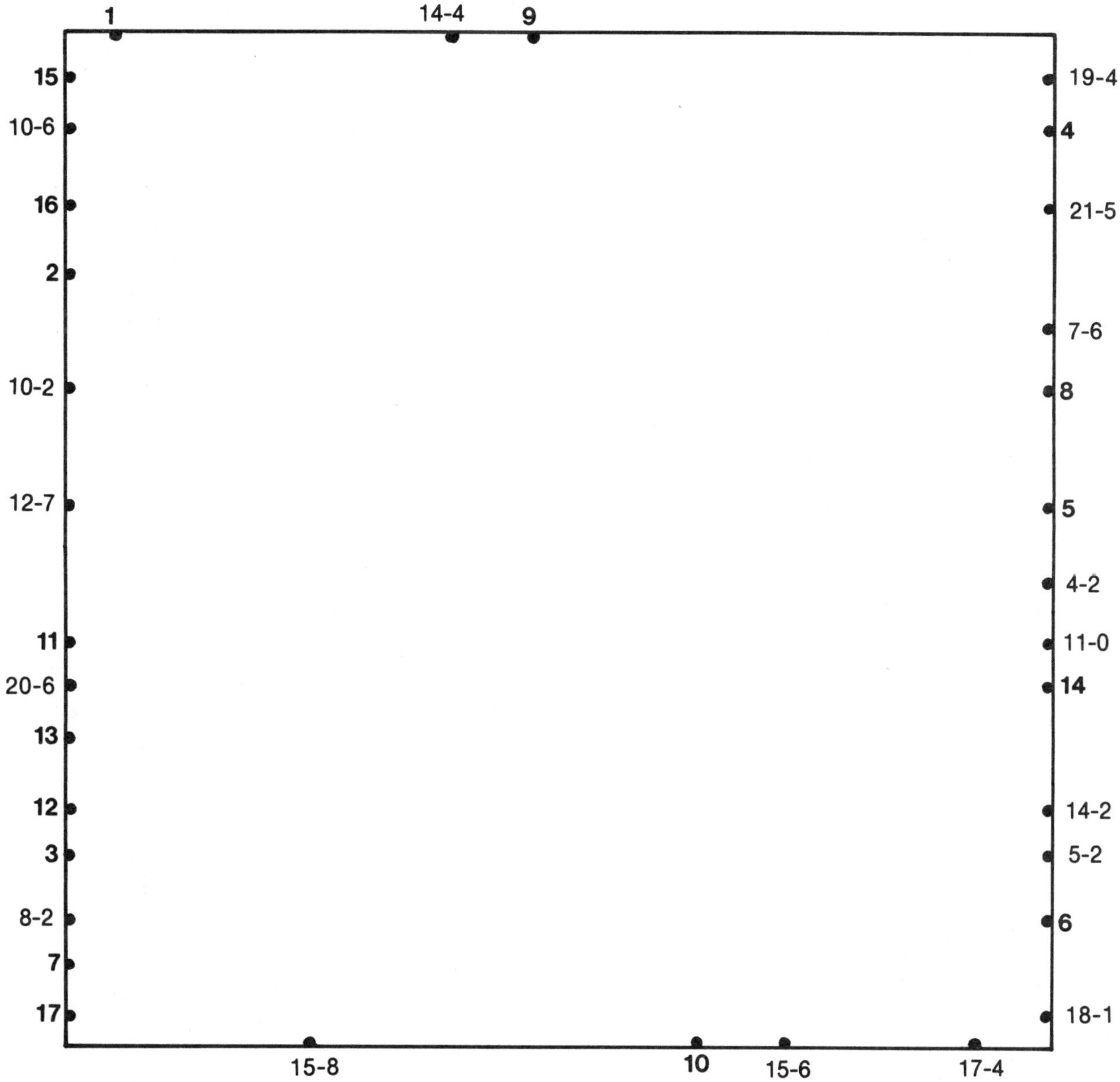

Correction Key
Design #3

1. Use this design as a correction key.
2. Allow your students to correct their own work.
3. Make a transparency of this design and instruct your students to place the transparency over their completed design for a quick and easy check.

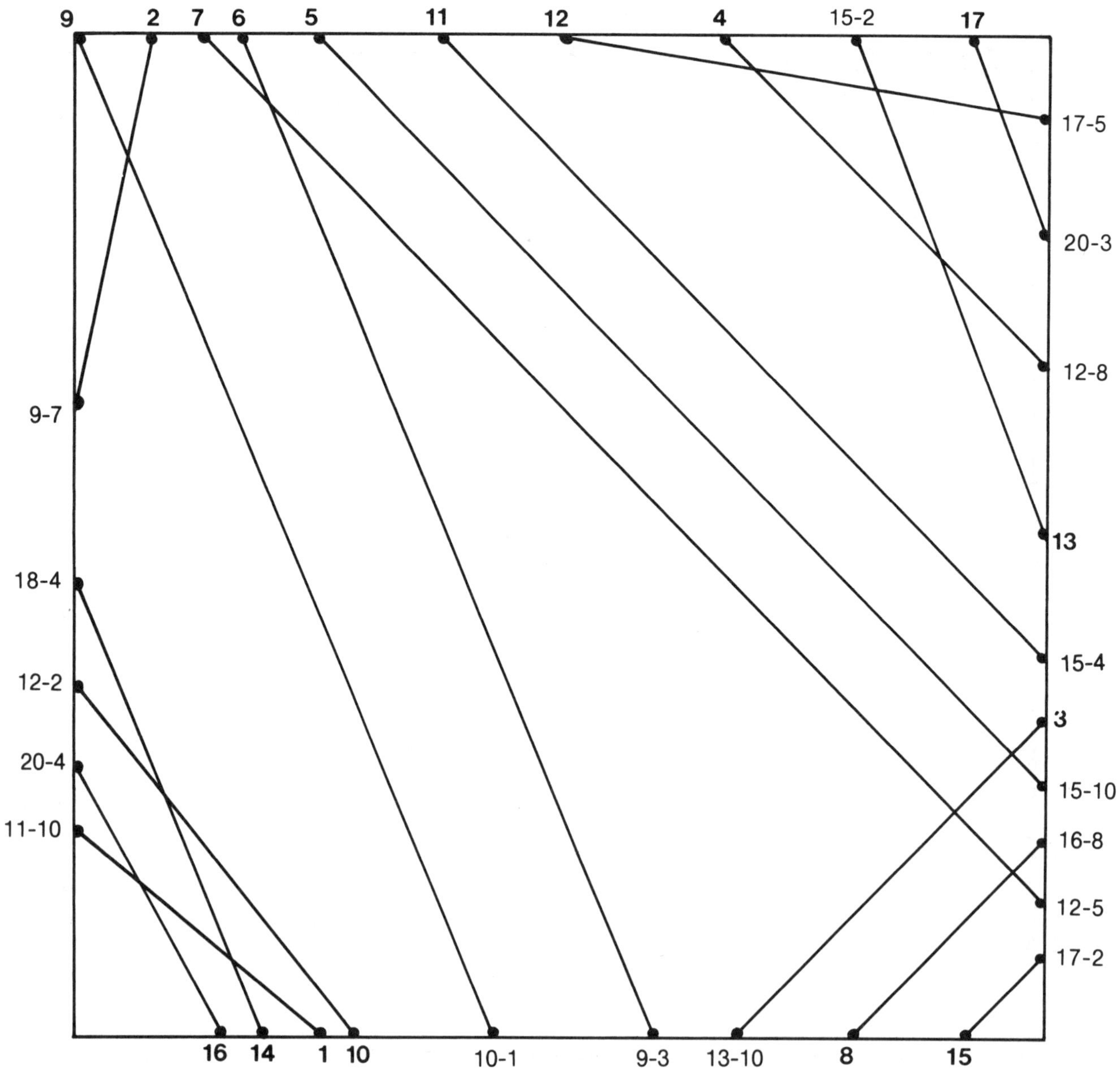

Subtraction
Design #3

Name ____________________

Date ____________________

1. **Read all of the directions before you begin.**
2. **Complete one subtraction problem next to a dot.**
   (You may use another piece of paper if you need to.)
3. **After you complete the problem, find the answer on the paper.**
4. **Use a ruler (or straight edge) and draw a line to connect the dot by the problem to the dot by the correct answer.**
5. **Complete each of the problems in the same way.**
6. **Carefully color your design after completing *all* of the problems.**

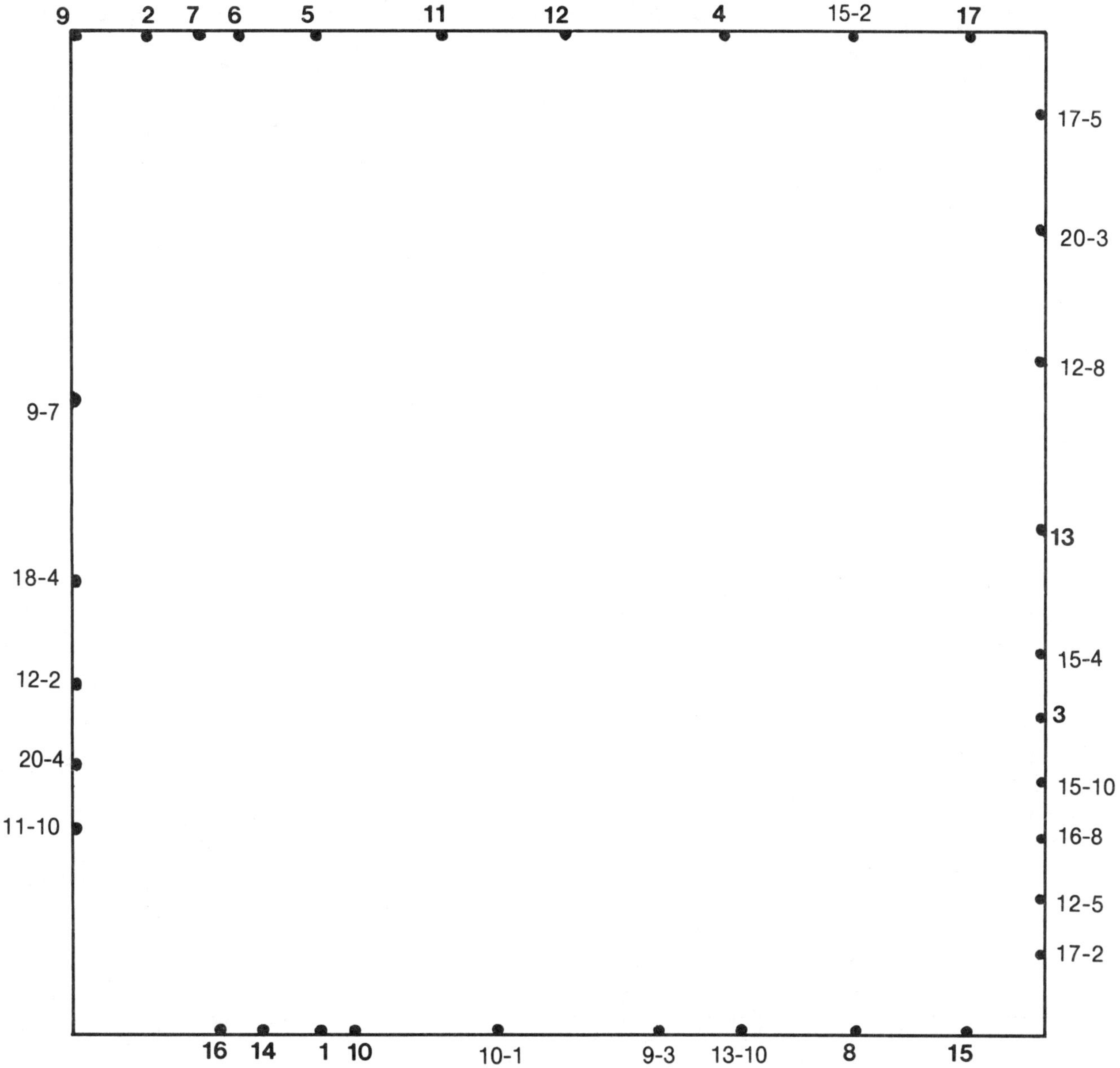

Correction Key
Design #4

1. Use this design as a correction key.
2. Allow your students to correct their own work.
3. Make a transparency of this design and instruct your students to place the transparency over their completed design for a quick and easy check.

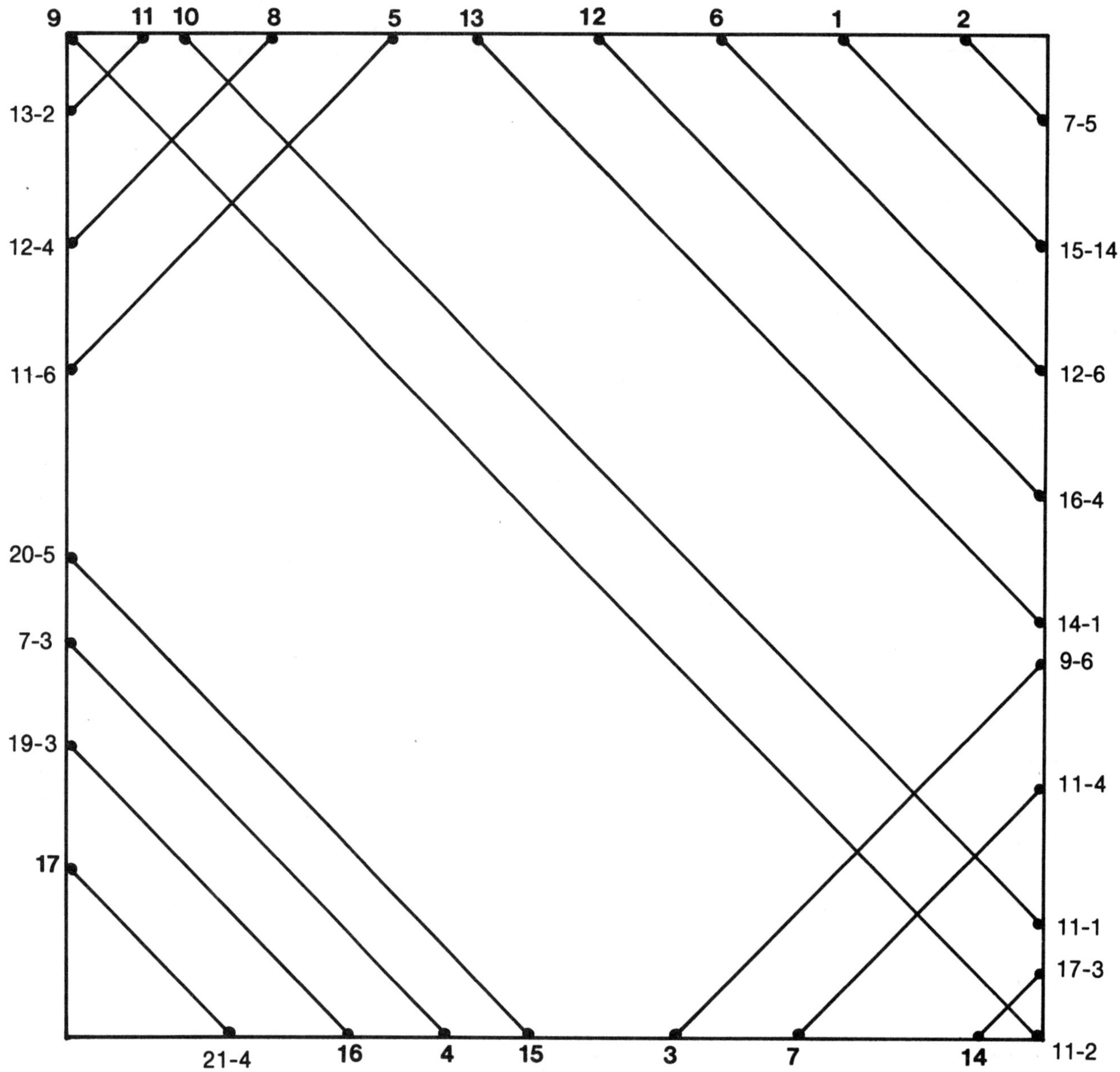

Subtraction

Design #4

Name ______________________

Date ______________________

1. Read all of the directions before you begin.
2. Complete one subtraction problem next to a dot.
   (You may use another piece of paper if you need to.)
3. After you complete the problem, find the answer on the paper.
4. Use a ruler (or straight edge) and draw a line to connect the dot by the problem to the dot by the correct answer.
5. Complete each of the problems in the same way.
6. Carefully color your design after completing *all* of the problems.

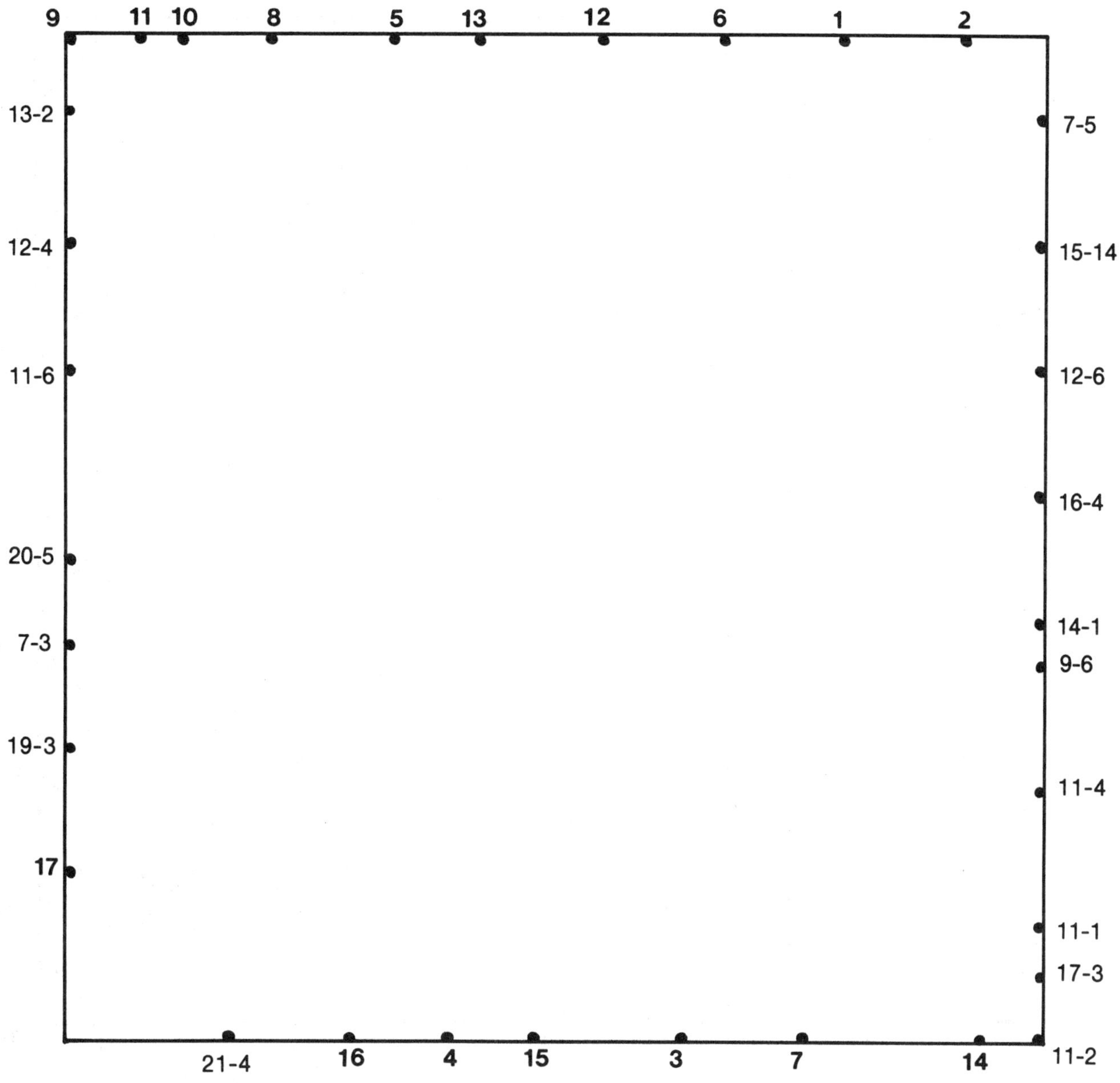

Correction Key
Design #5

1. Use this design as a correction key.
2. Allow your students to correct their own work.
3. Make a transparency of this design and instruct your students to place the transparency over their completed design for a quick and easy check.

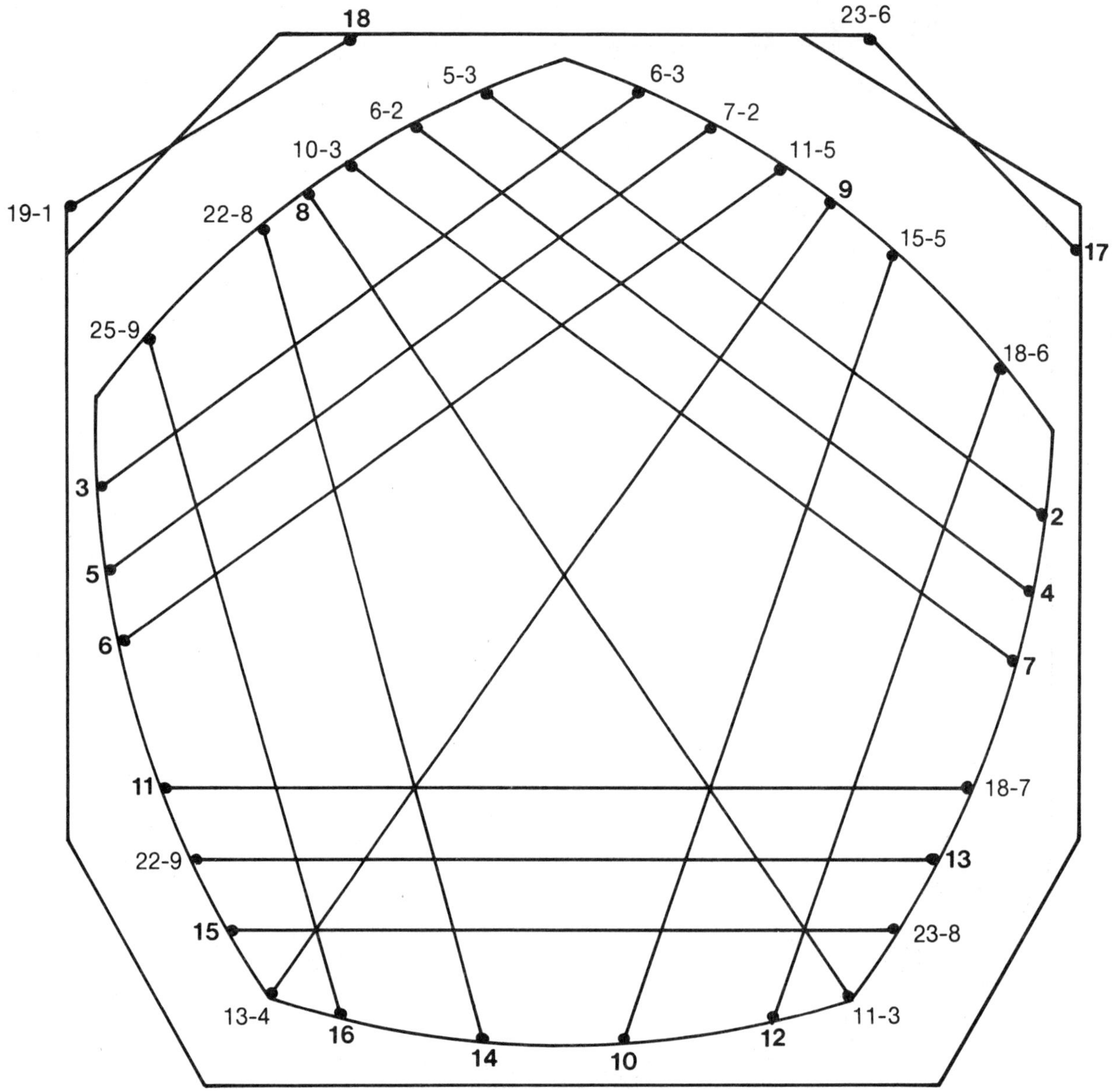

Subtraction

Design #5

Name ______________________

Date ______________________

1. **Read all of the directions before you begin.**
2. **Complete one subtraction problem next to a dot.**
   (You may use another piece of paper if you need to.)
3. **After you complete the problem, find the answer on the paper.**
4. **Use a ruler (or straight edge) and draw a line to connect the dot by the problem to the dot by the correct answer.**
5. **Complete each of the problems in the same way.**
6. **Carefully color your design after completing *all* of the problems.**

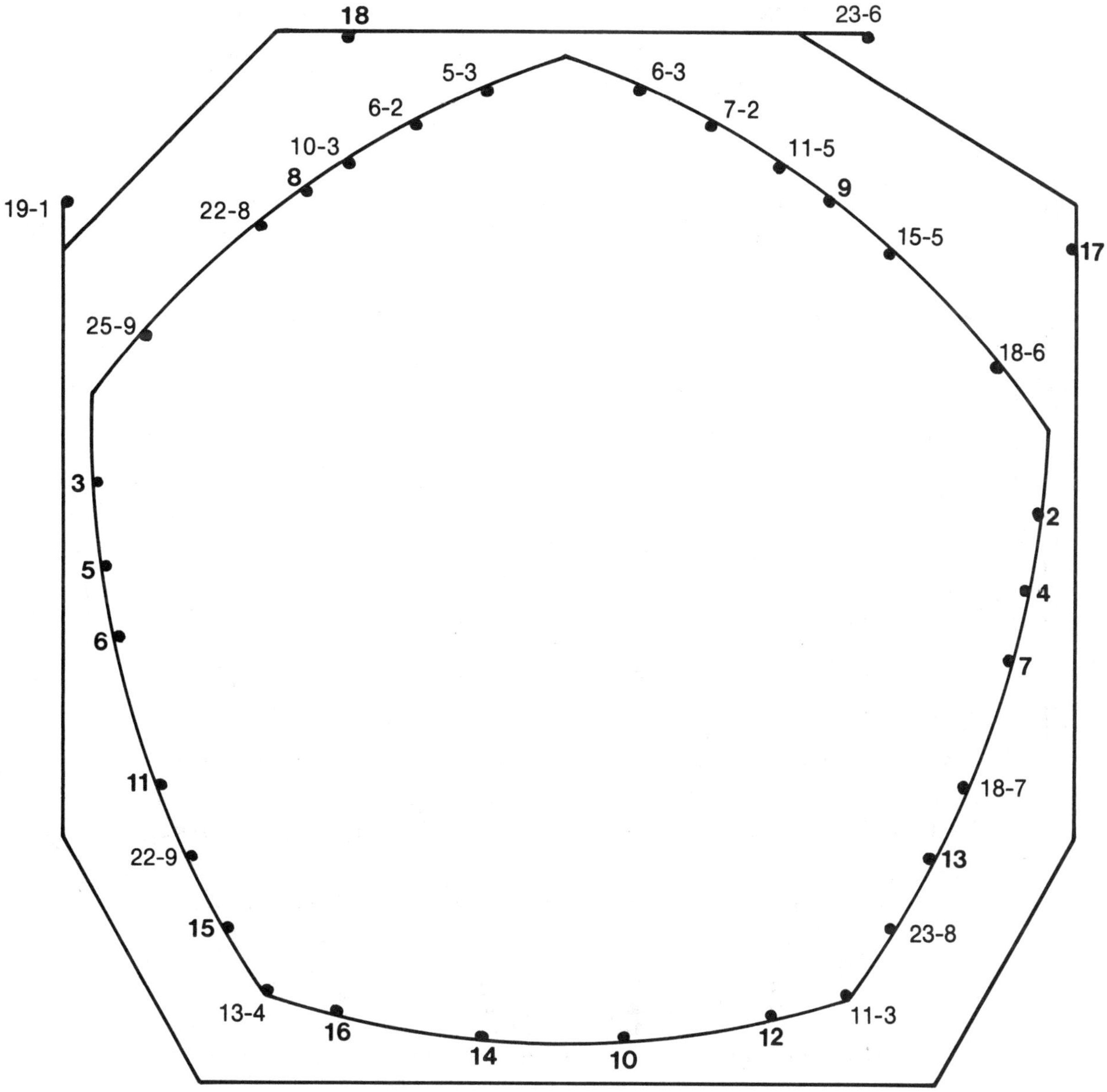

Correction Key
Design #6

1. Use this design as a correction key.
2. Allow your students to correct their own work.
3. Make a transparency of this design and instruct your students to place the transparency over their completed design for a quick and easy check.

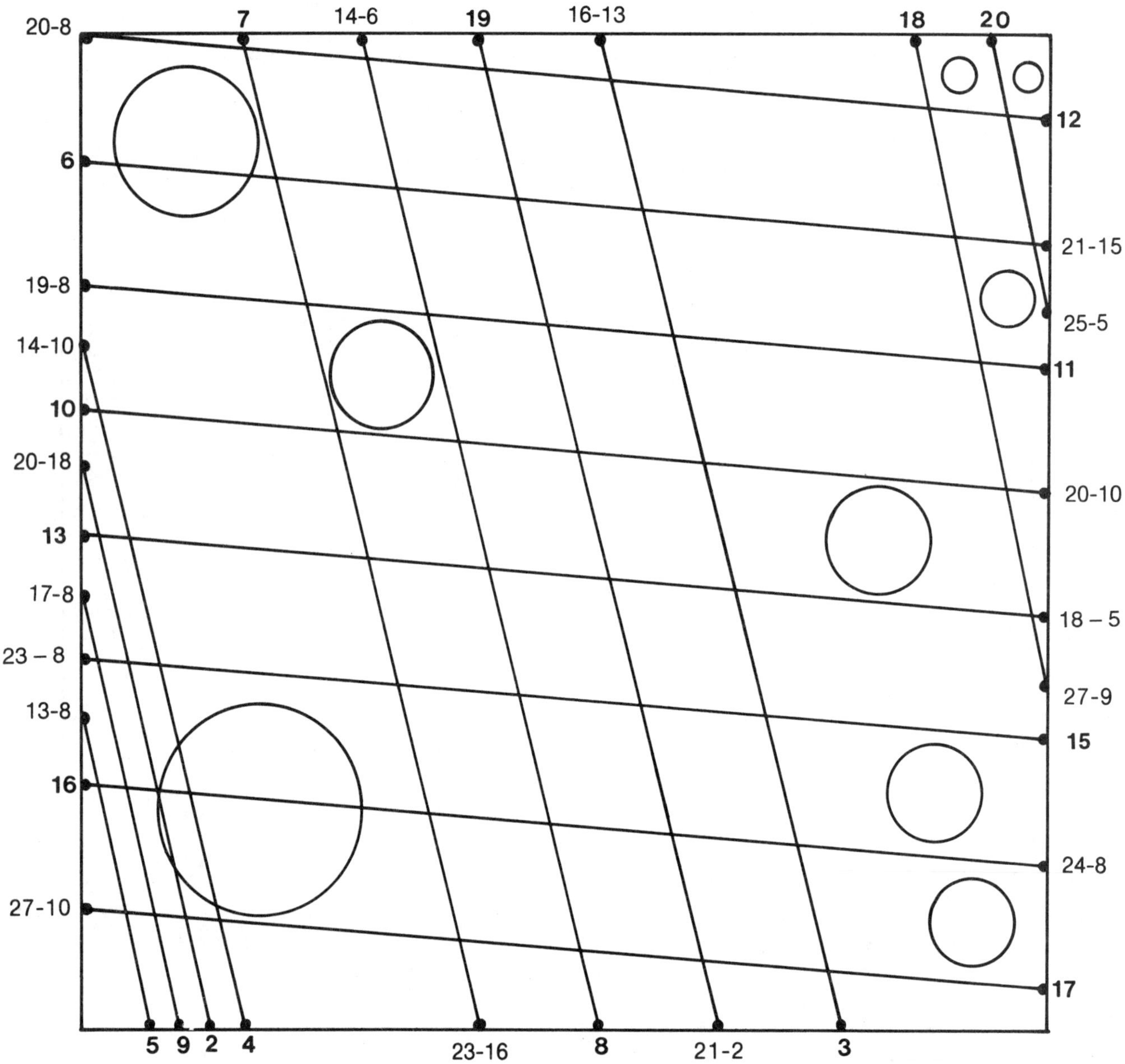

Subtraction
Design #6

Name ______________________

Date ______________________

1. Read all of the directions before you begin.
2. Complete one subtraction problem next to a dot.
   (You may use another piece of paper if you need to.)
3. After you complete the problem, find the answer on the paper.
4. Use a ruler (or straight edge) and draw a line to connect the dot by the problem to the dot by the correct answer.
5. Complete each of the problems in the same way.
6. Carefully color your design after completing *all* of the problems.

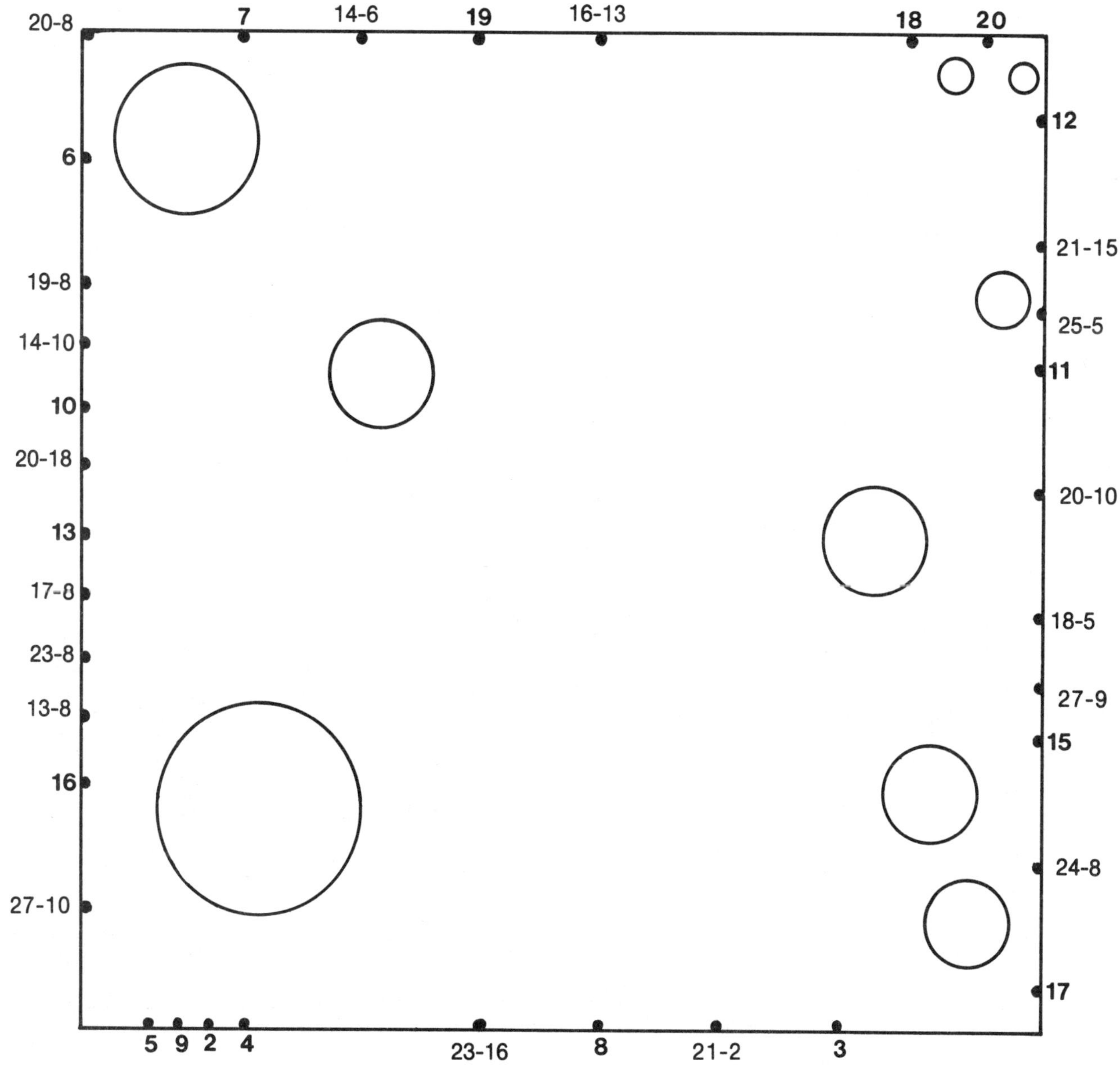

Correction Key
Design #7

1. Use this design as a correction key.
2. Allow your students to correct their own work.
3. Make a transparency of this design and instruct your students to place the transparency over their completed design for a quick and easy check.

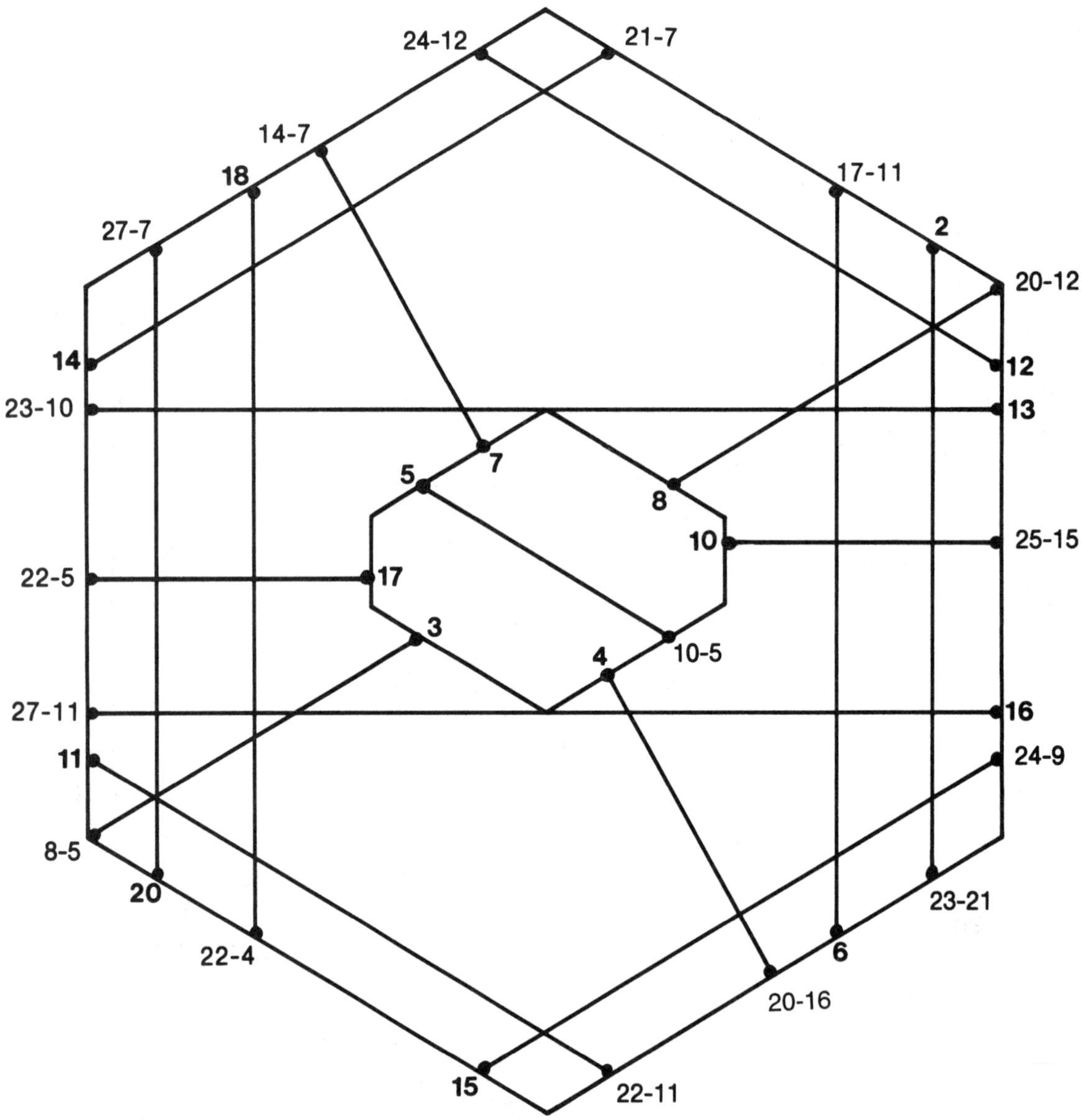

Subtraction
Design #7

Name ____________________

Date ____________________

1. **Read all of the directions before you begin.**
2. **Complete one subtraction problem next to a dot.**
   (You may use another piece of paper if you need to.)
3. **After you complete the problem, find the answer on the paper.**
4. **Use a ruler (or straight edge) and draw a line to connect the dot by the problem to the dot by the correct answer.**
5. **Complete each of the problems in the same way.**
6. **Carefully color your design after completing *all* of the problems.**

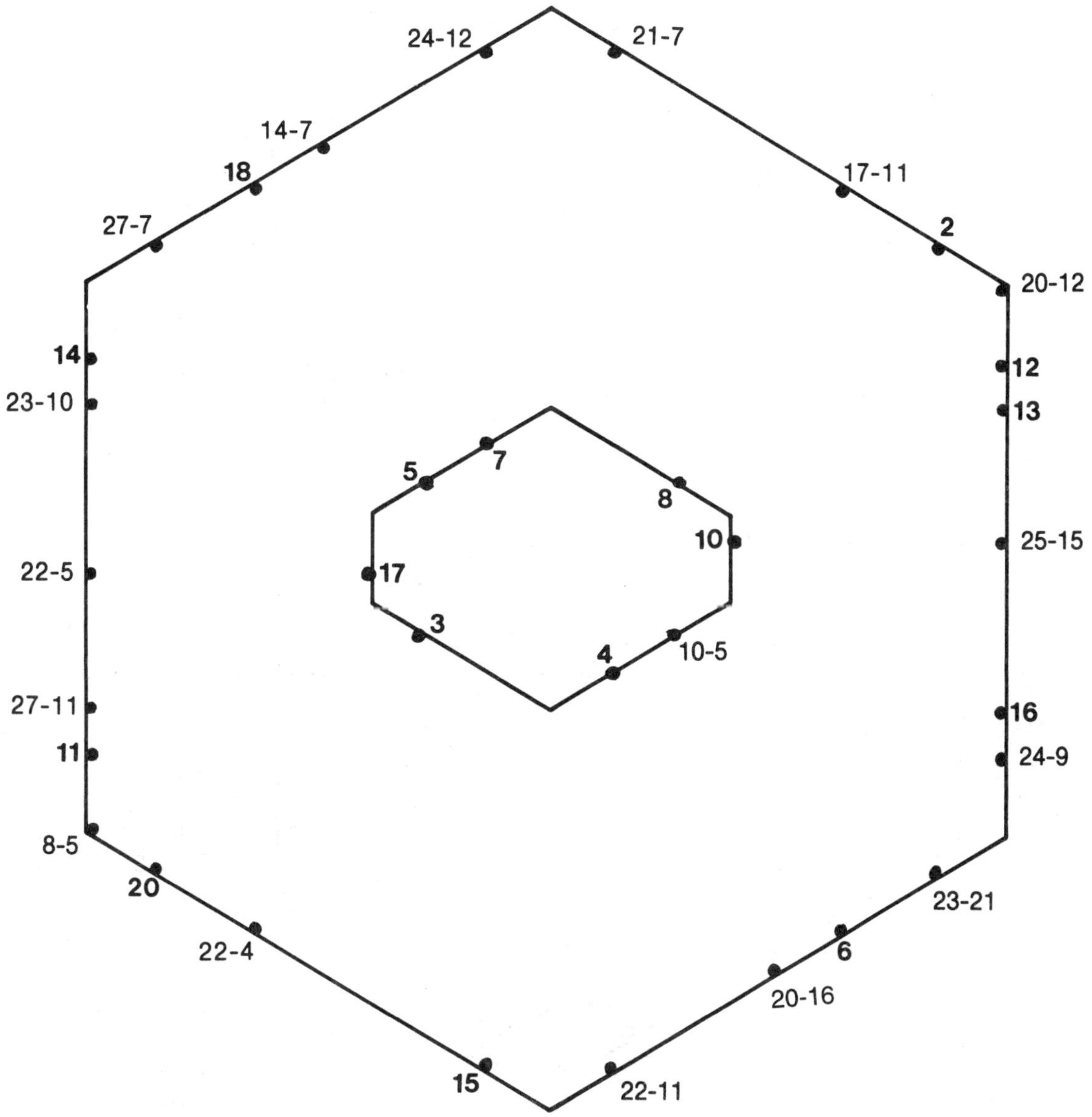

Correction Key
Design #8

1. Use this design as a correction key.
2. Allow your students to correct their own work.
3. Make a transparency of this design and instruct your students to place the transparency over their completed design for a quick and easy check.

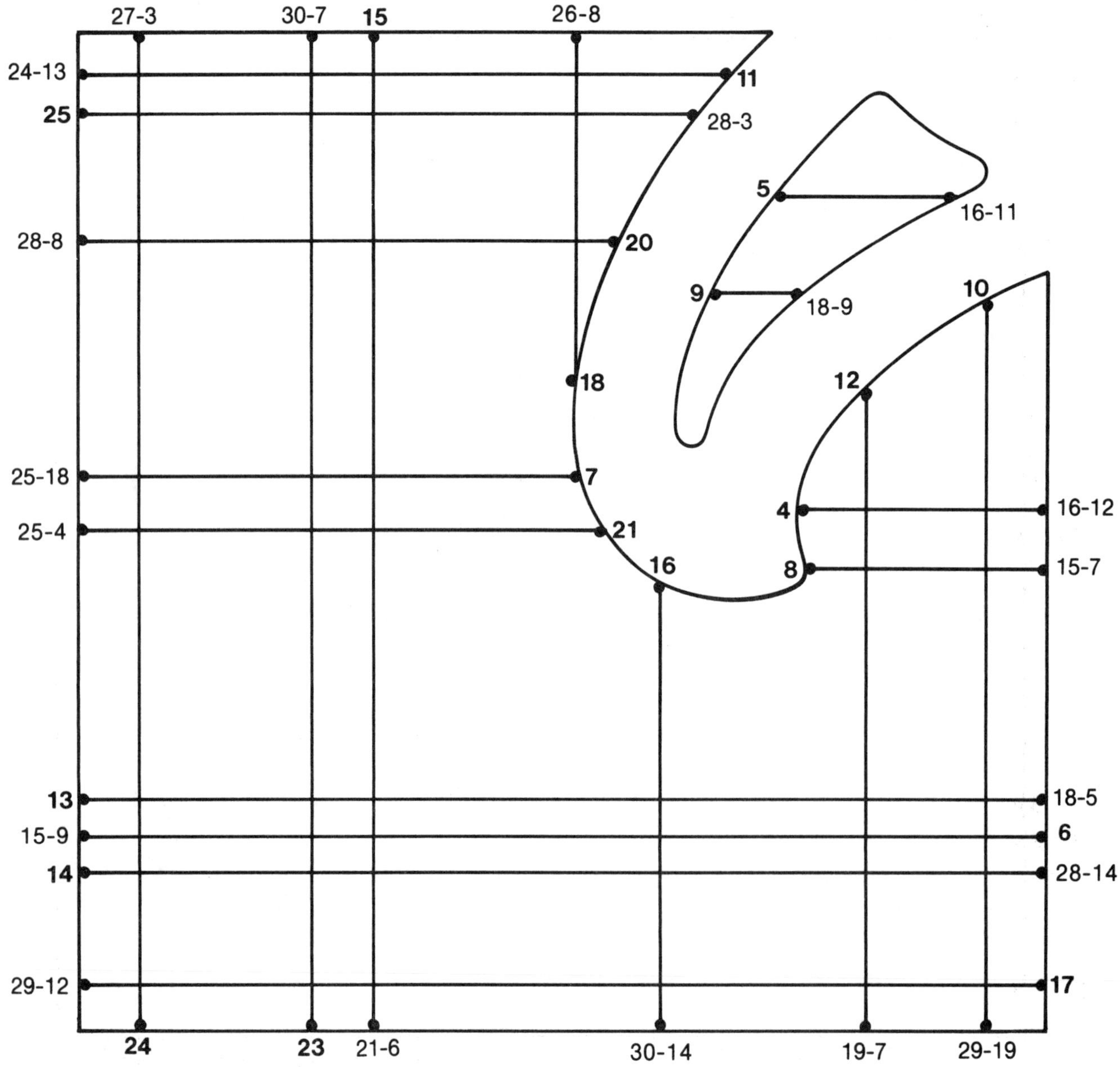

Subtraction
Design #8

Name ______________________

Date ______________________

1. **Read all of the directions before you begin.**
2. **Complete one subtraction problem next to a dot.**
   (You may use another piece of paper if you need to.)
3. **After you complete the problem, find the answer on the paper.**
4. **Use a ruler (or straight edge) and draw a line to connect the dot by the problem to the dot by the correct answer.**
5. **Complete each of the problems in the same way.**
6. **Carefully color your design after completing *all* of the problems.**

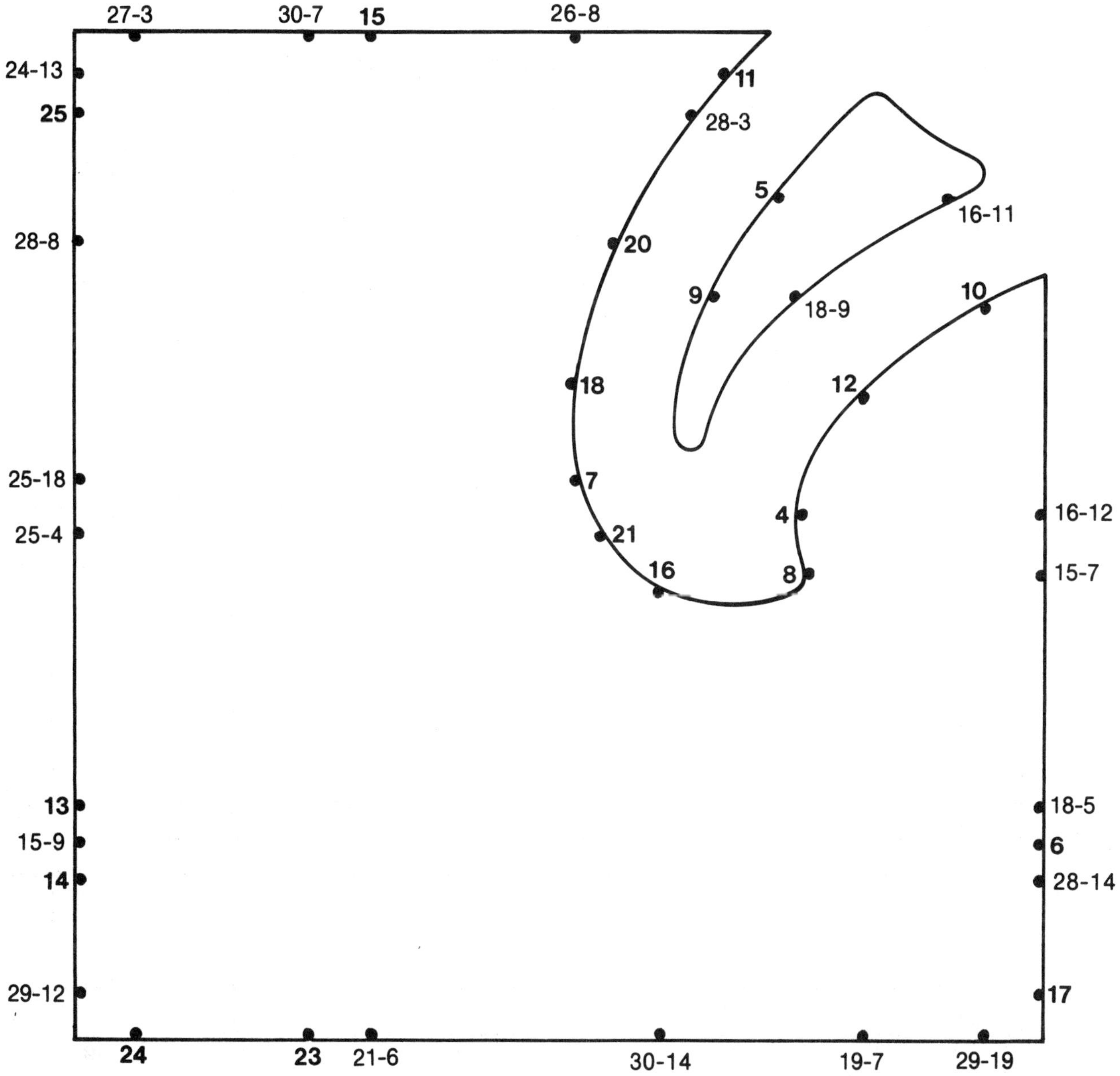

Correction Key
Design #9

1. Use this design as a correction key.
2. Allow your students to correct their own work.
3. Make a transparency of this design and instruct your students to place the transparency over their completed design for a quick and easy check.

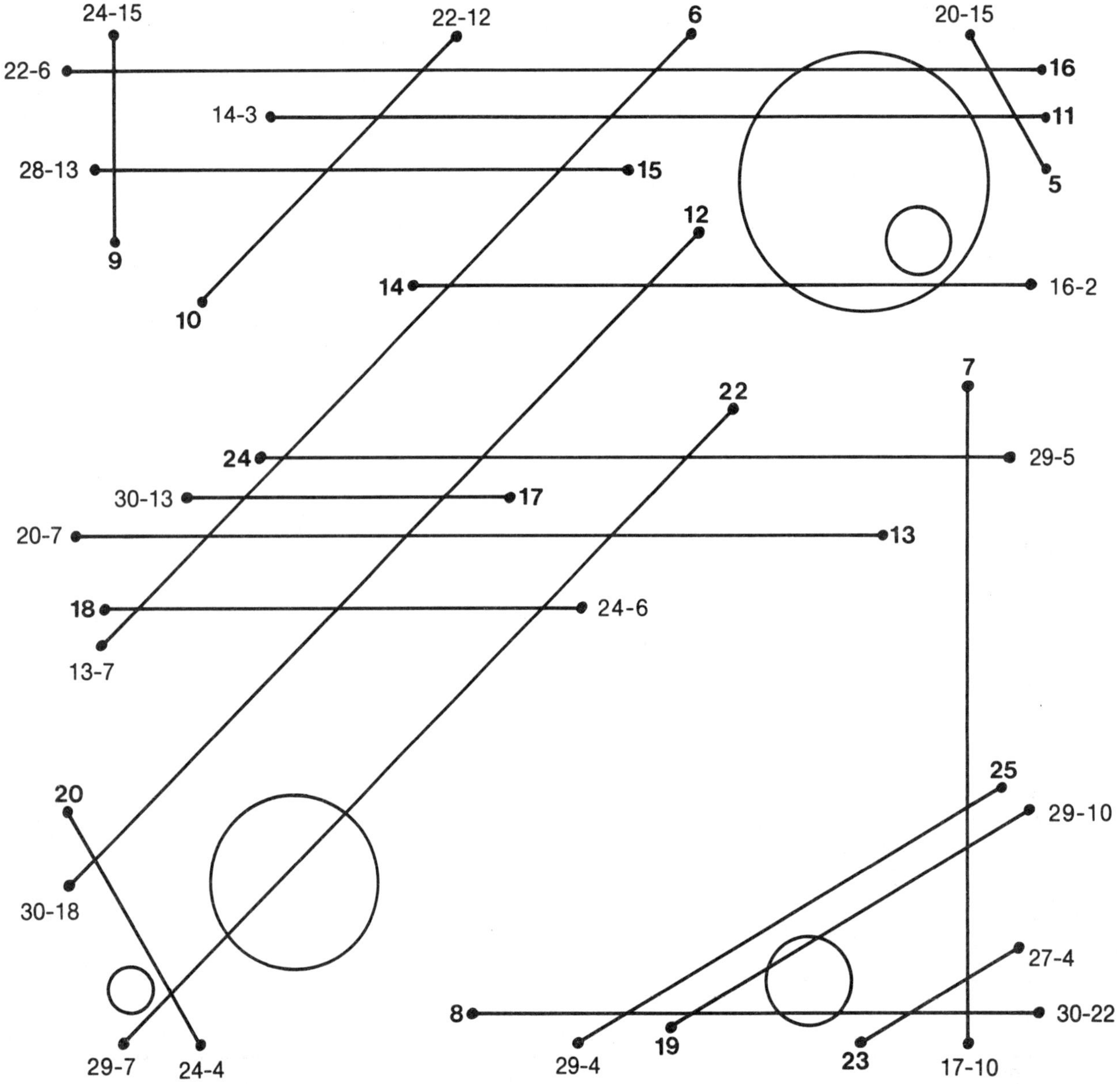

Subtraction
Design #9

Name ______________________________

Date ______________________________

1. **Read all of the directions before you begin.**
2. **Complete one subtraction problem next to a dot.**
   (You may use another piece of paper if you need to.)
3. **After you complete the problem, find the answer on the paper.**
4. **Use a ruler (or straight edge) and draw a line to connect the dot by the problem to the dot by the correct answer.**
5. **Complete each of the problems in the same way.**
6. **Carefully color your design after completing *all* of the problems.**

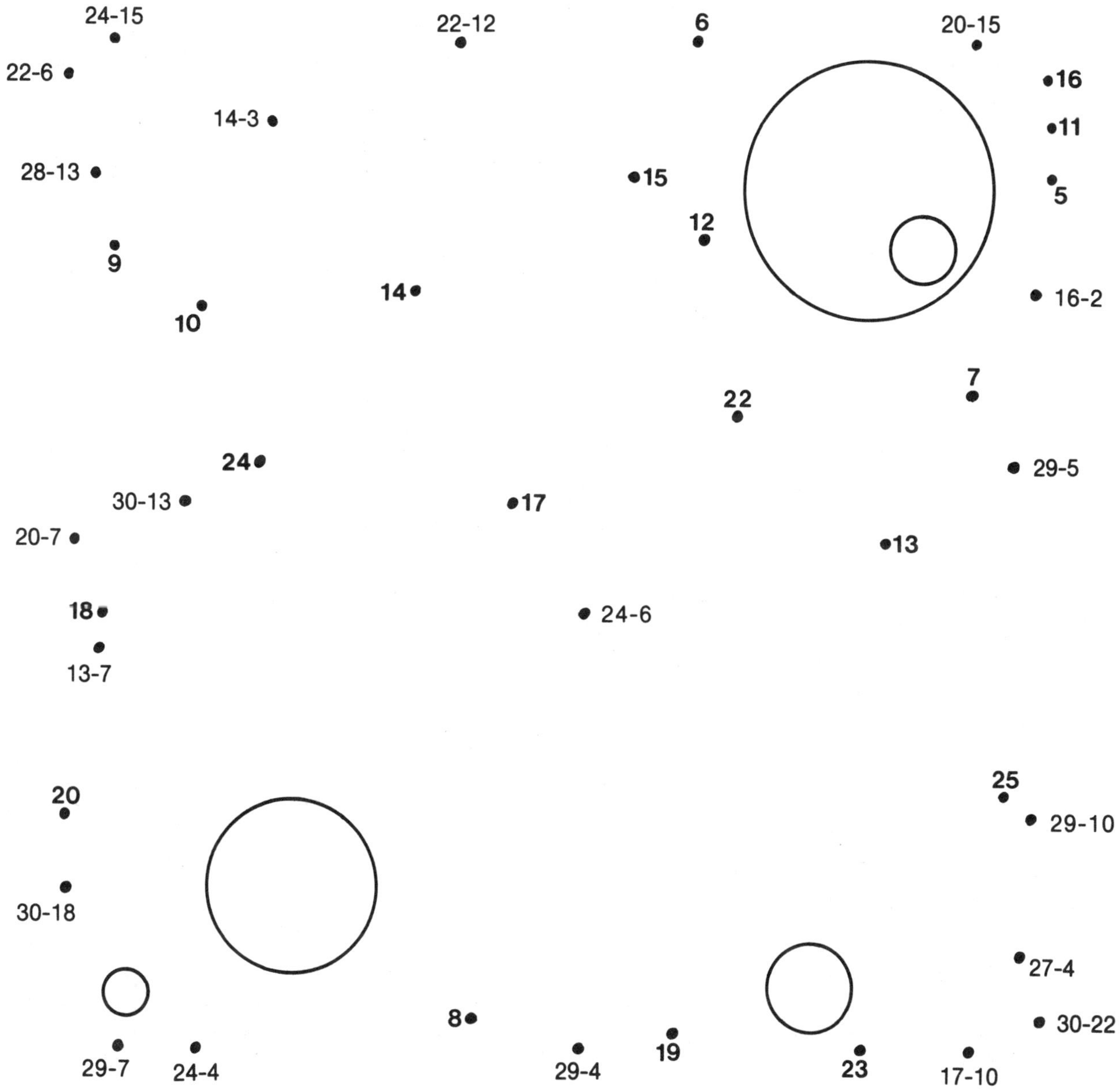

Correction Key
Design #10

1. Use this design as a correction key.
2. Allow your students to correct their own work.
3. Make a transparency of this design and instruct your students to place the transparency over their completed design for a quick and easy check.

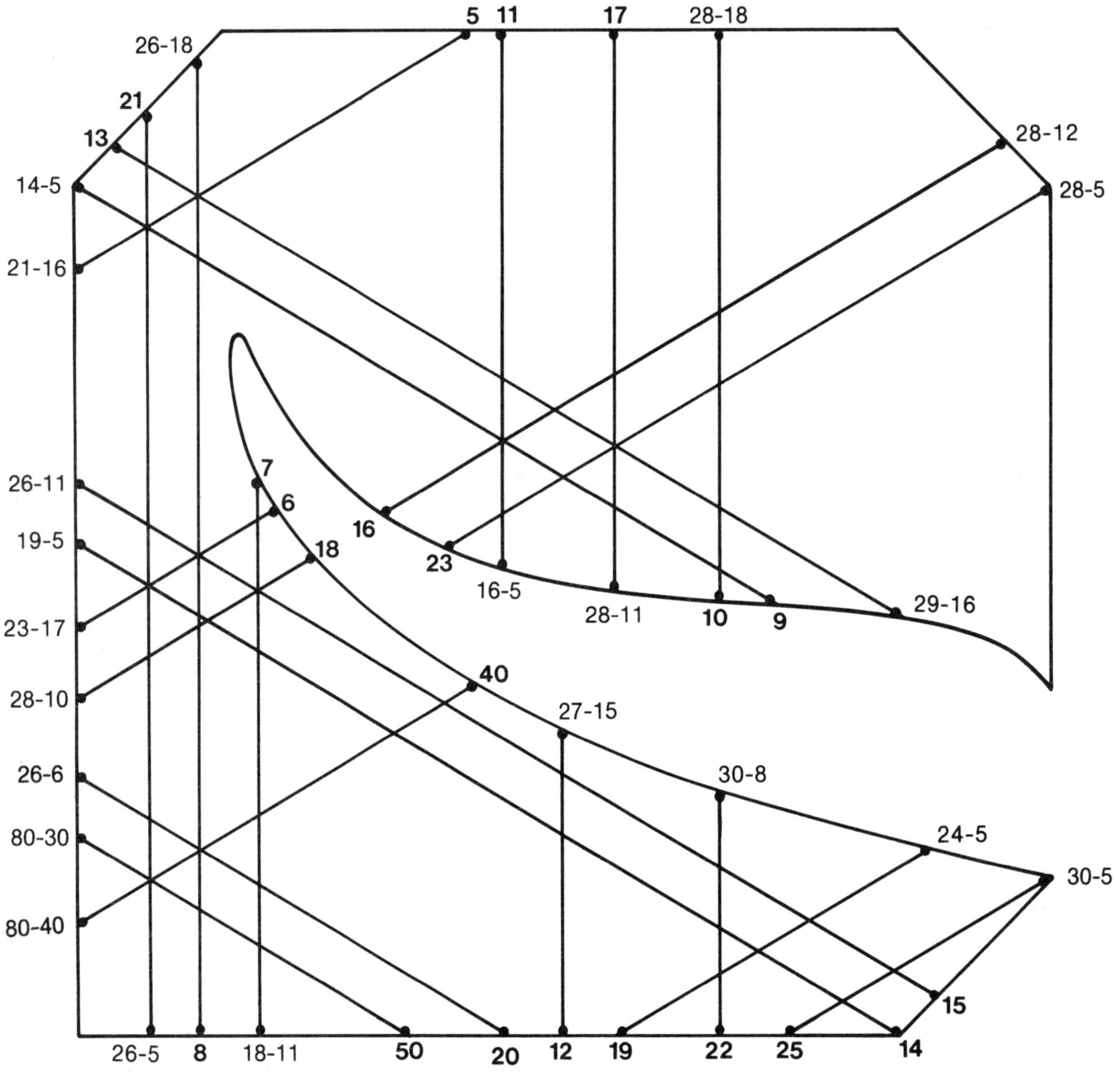

Subtraction
Design #10

Name ______________________

Date ______________________

1. **Read all of the directions before you begin.**
2. **Complete one subtraction problem next to a dot.** (You can use another piece of paper if you need to.)
3. **After you complete the problem, find the answer on the paper.**
4. **Use a ruler (or straight edge) and draw a line to connect the dot by the problem to the dot by the correct answer.**
5. **Complete each of the problems in the same way.**
6. **Carefully color your design after completing *all* of the problems.**

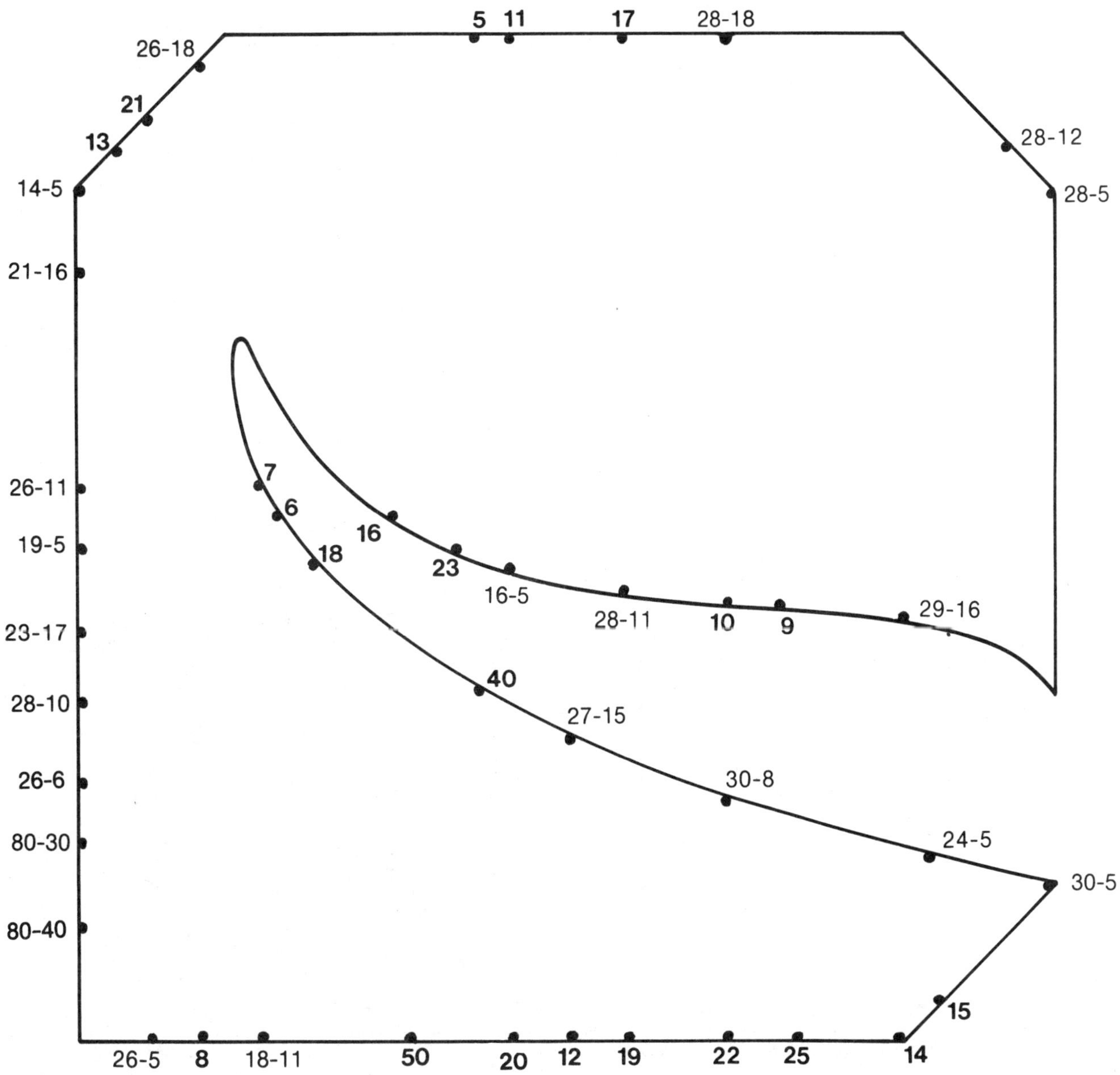

Correction Key
Design #11

1. Use this design as a correction key.
2. Allow your students to correct their own work.
3. Make a transparency of this design and instruct your students to place the transparency over their completed design for a quick and easy check.

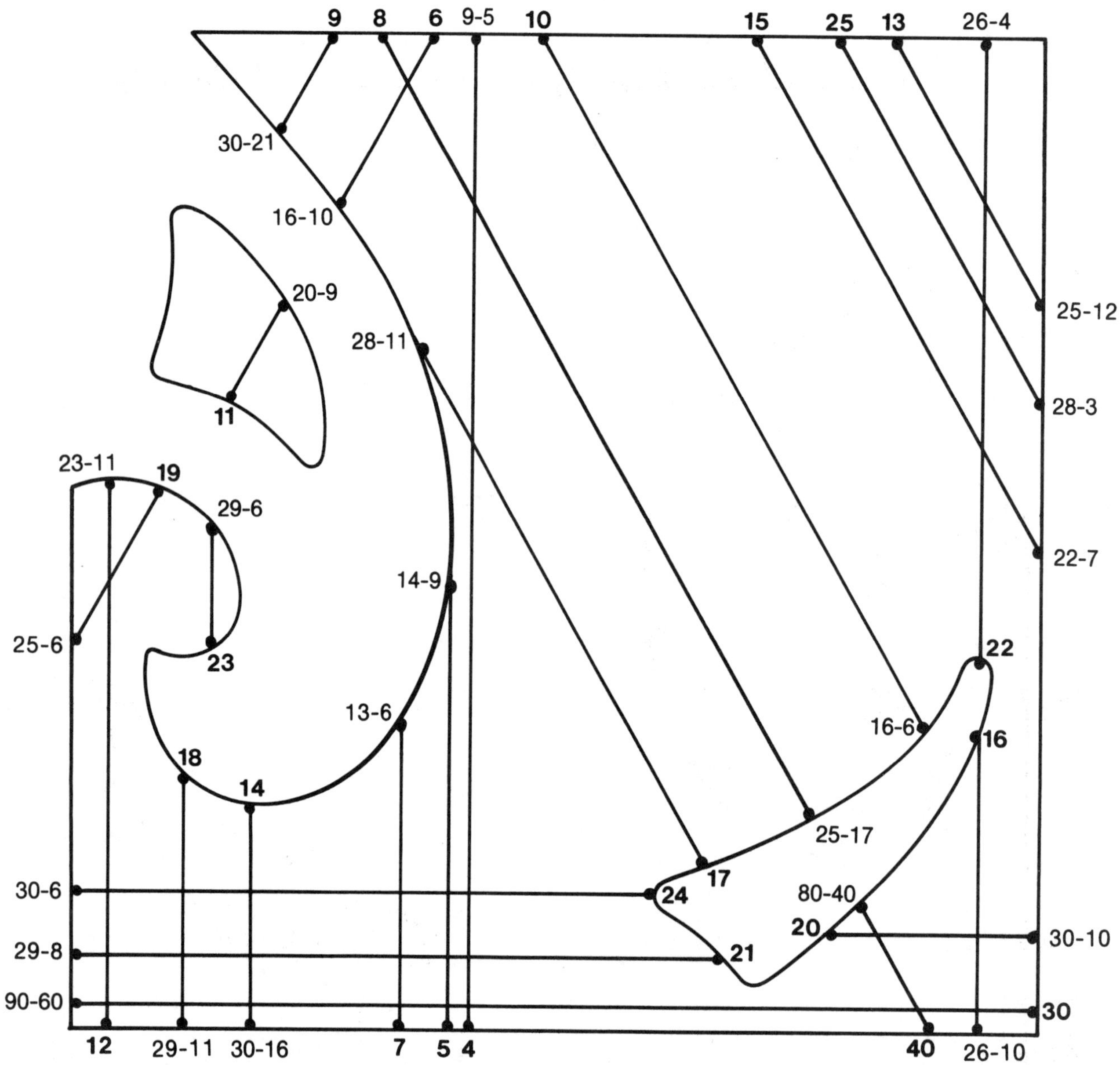

Subtraction

Design #11

Name ______________________________

Date ______________________________

1. Read all of the directions before you begin.
2. Complete one subtraction problem next to a dot.
   (You can use another piece of paper if you need to.)
3. After you complete the problem, find the answer on the paper.
4. Use a ruler (or straight edge) and draw a line to connect the dot by the problem to the dot by the correct answer.
5. Complete each of the problems in the same way.
6. Carefully color your design after completing *all* of the problems.

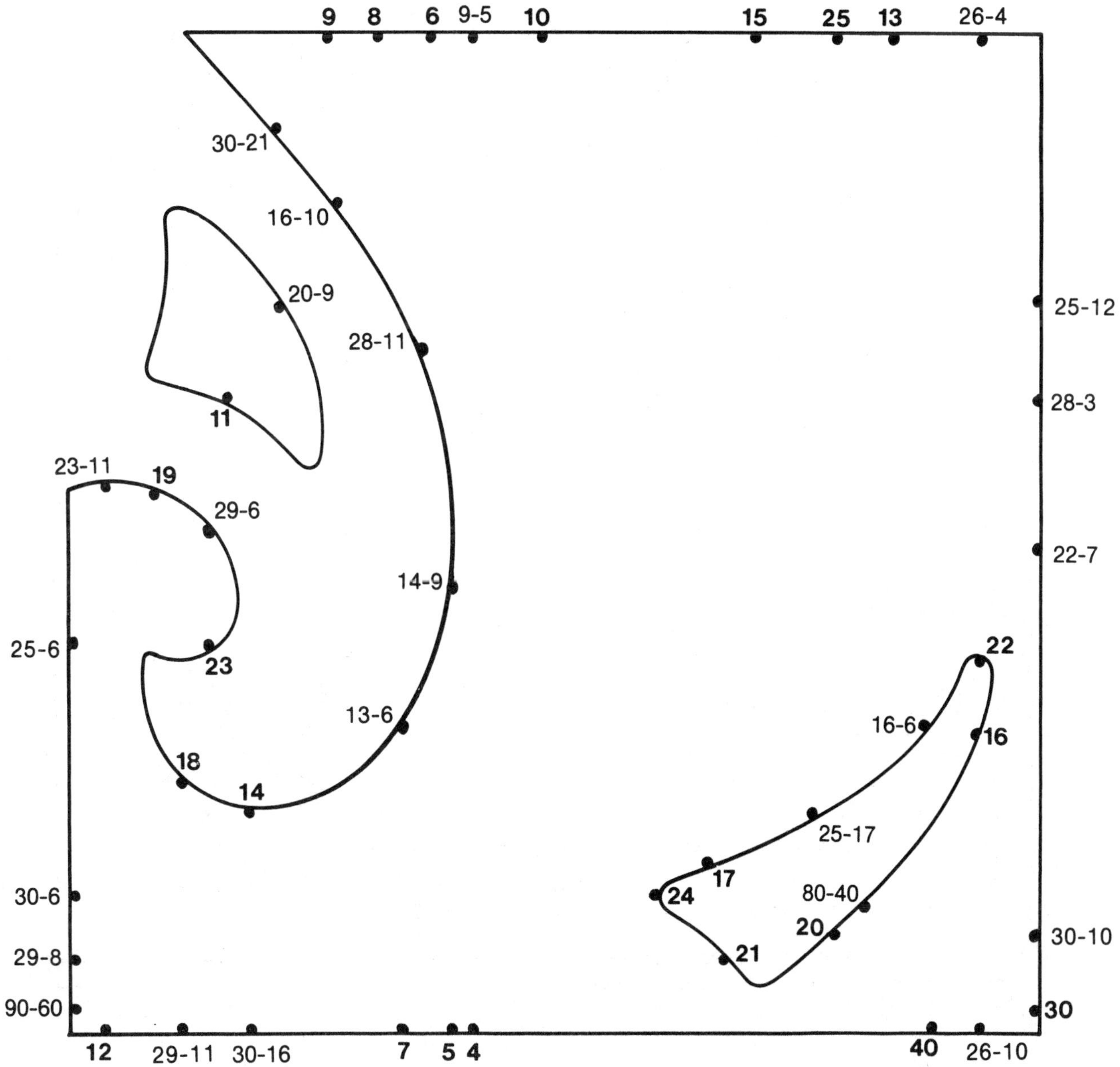

Correction Key
Design #12

1. Use this design as a correction key.
2. Allow your students to correct their own work.
3. Make a transparency of this design and instruct your students to place the transparency over their completed design for a quick and easy check.

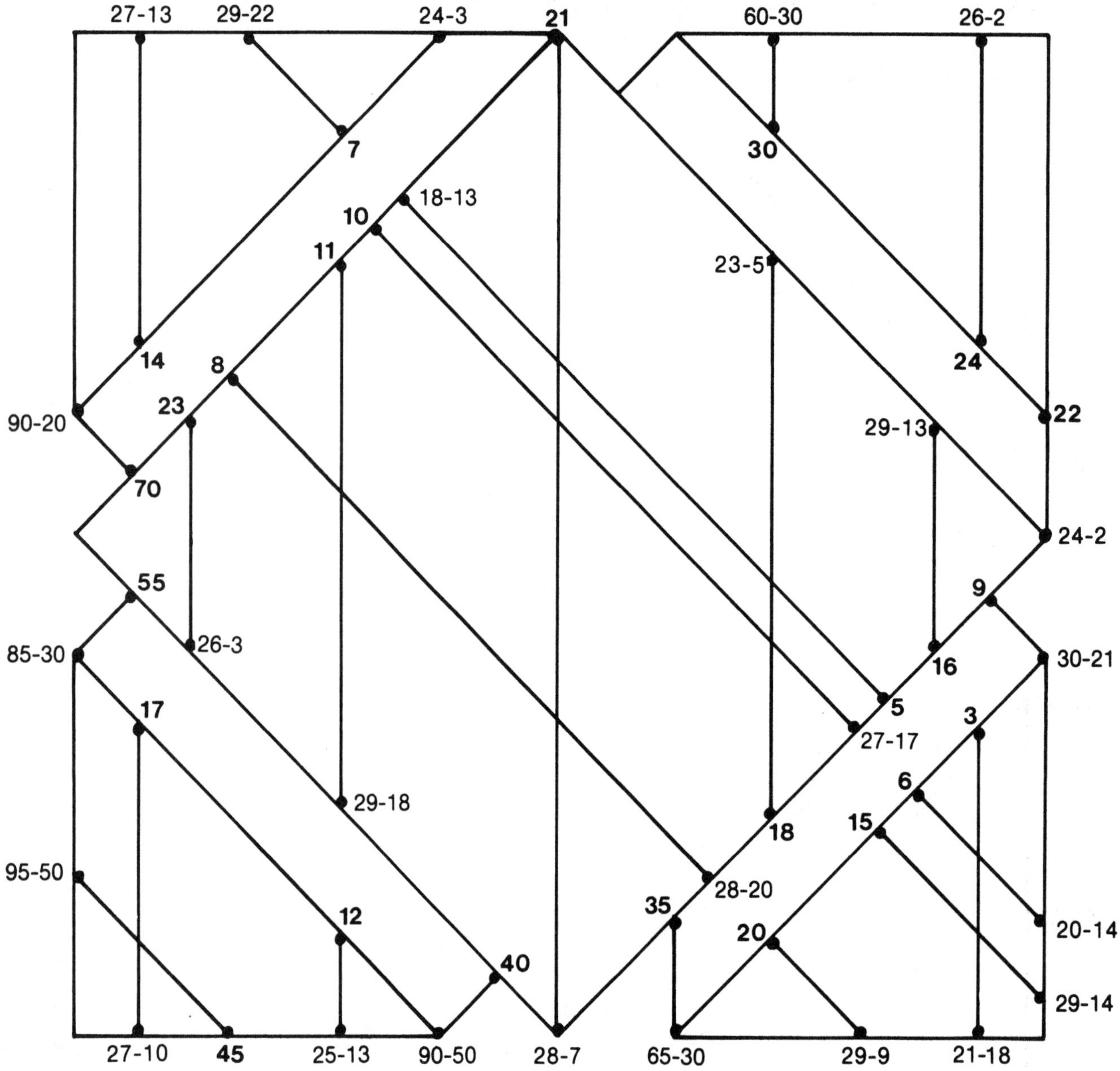

Subtraction
Design #12

Name ____________________

Date ____________________

1. Read all of the directions before you begin.
2. Complete one subtraction problem next to a dot. (You can use another piece of paper if you need to.)
3. After you complete the problem, find the answer on the paper.
4. Use a ruler (or straight edge) and draw a line to connect the dot by the problem to the dot by the correct answer.
5. Complete each of the problems in the same way.
6. Make sure that you complete all of the problems. Some answers may be used more than once.
7. Carefully color your design after completing *all* of the problems.

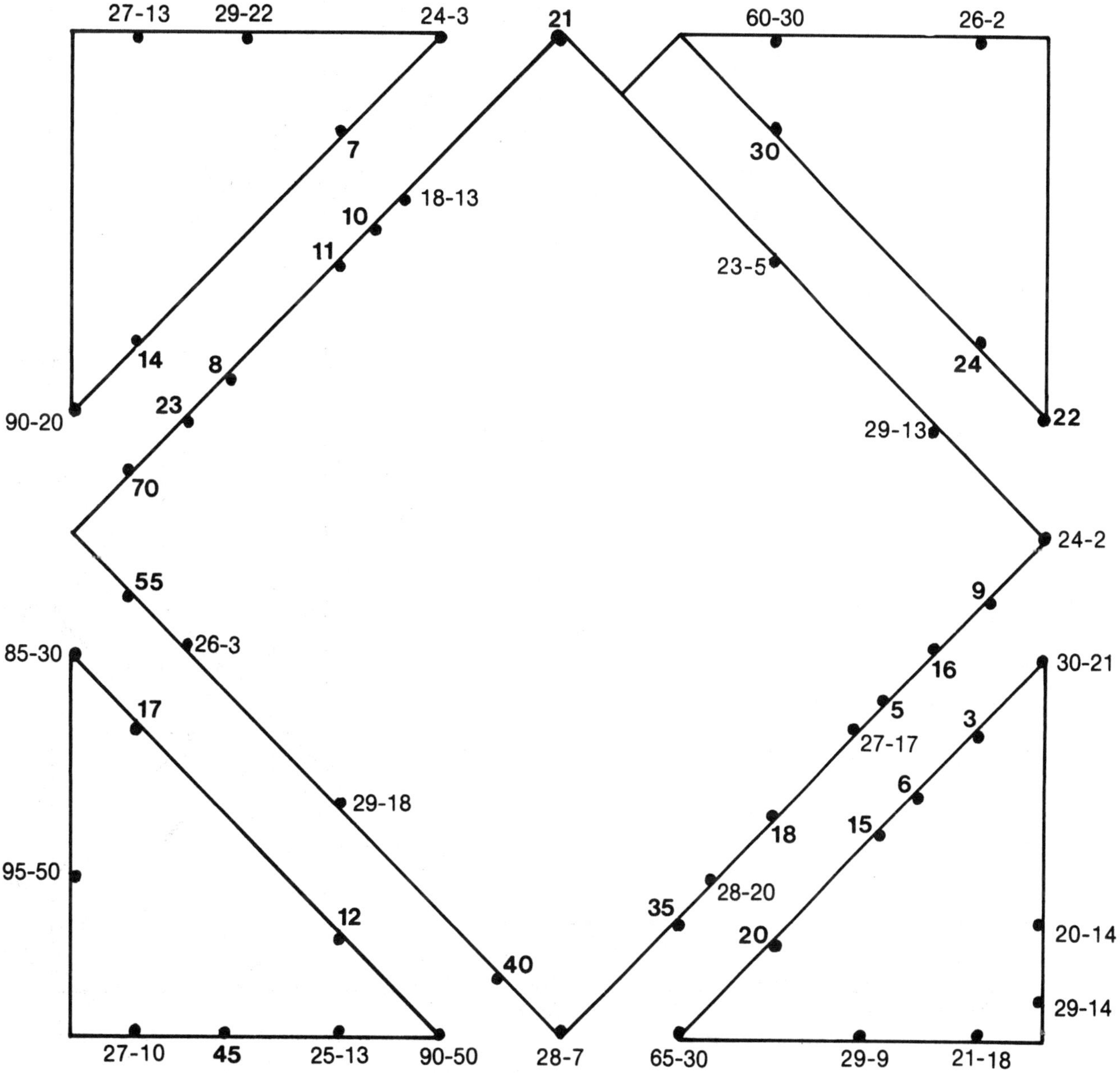

Correction Key
Design #13

1. Use this design as a correction key.
2. Allow your students to correct their own work.
3. Make a transparency of this design and instruct your students to place the transparency over their completed design for a quick and easy check.

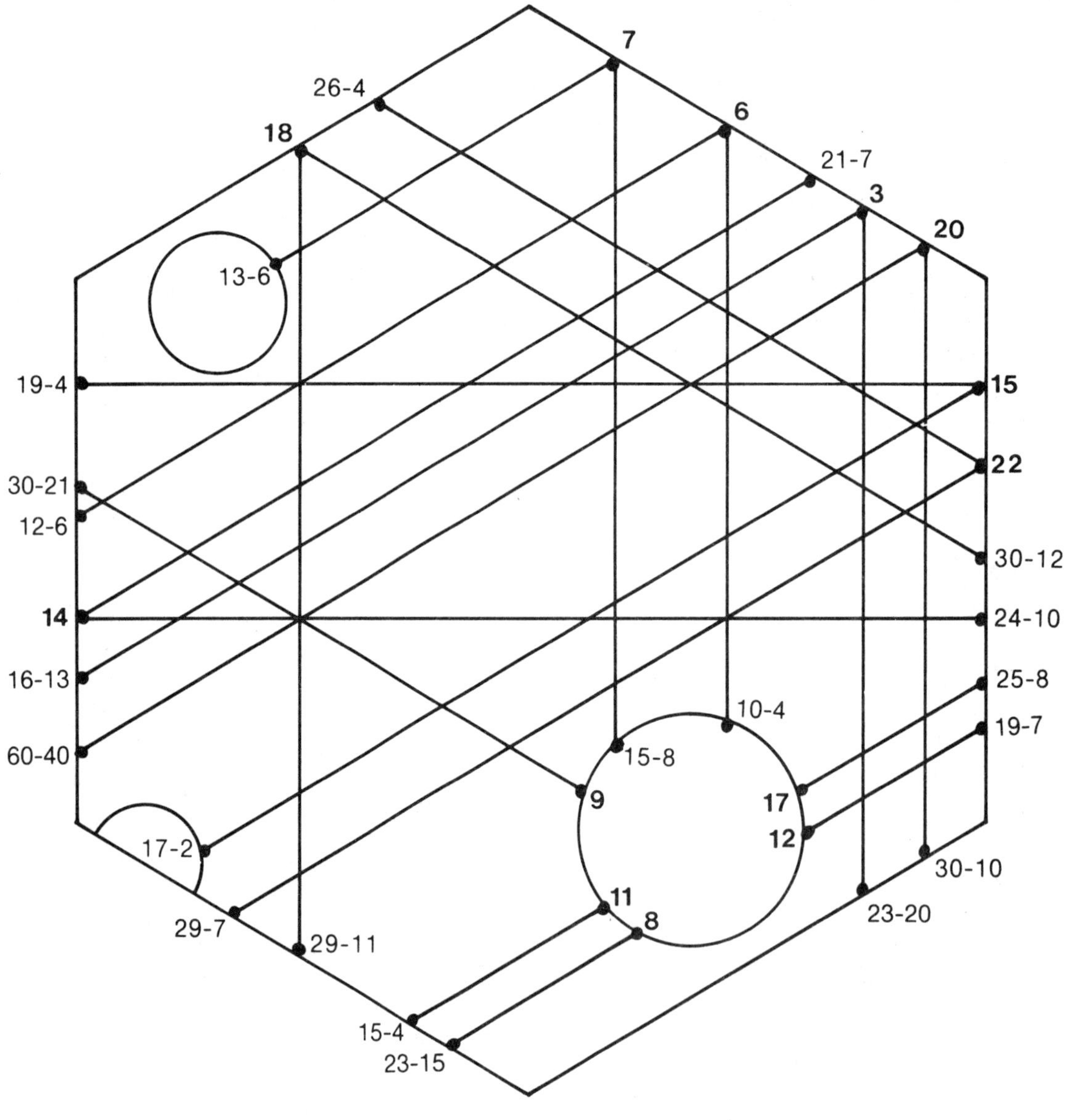

Subtraction
Design #13

Name ______________________

Date ______________________

1. **Read all of the directions before you begin.**
2. **Complete one subtraction problem next to a dot.** (You can use another piece of paper if you need to.)
3. **After you complete the problem, find the answer on the paper.**
4. **Use a ruler (or straight edge) and draw a line to connect the dot by the problem to the dot by the correct answer.**
5. **Complete each of the problems in the same way.**
6. **Make sure that you complete all of the problems. Some answers may be used more than once.**
7. **Carefully color your design after completing *all* of the problems.**

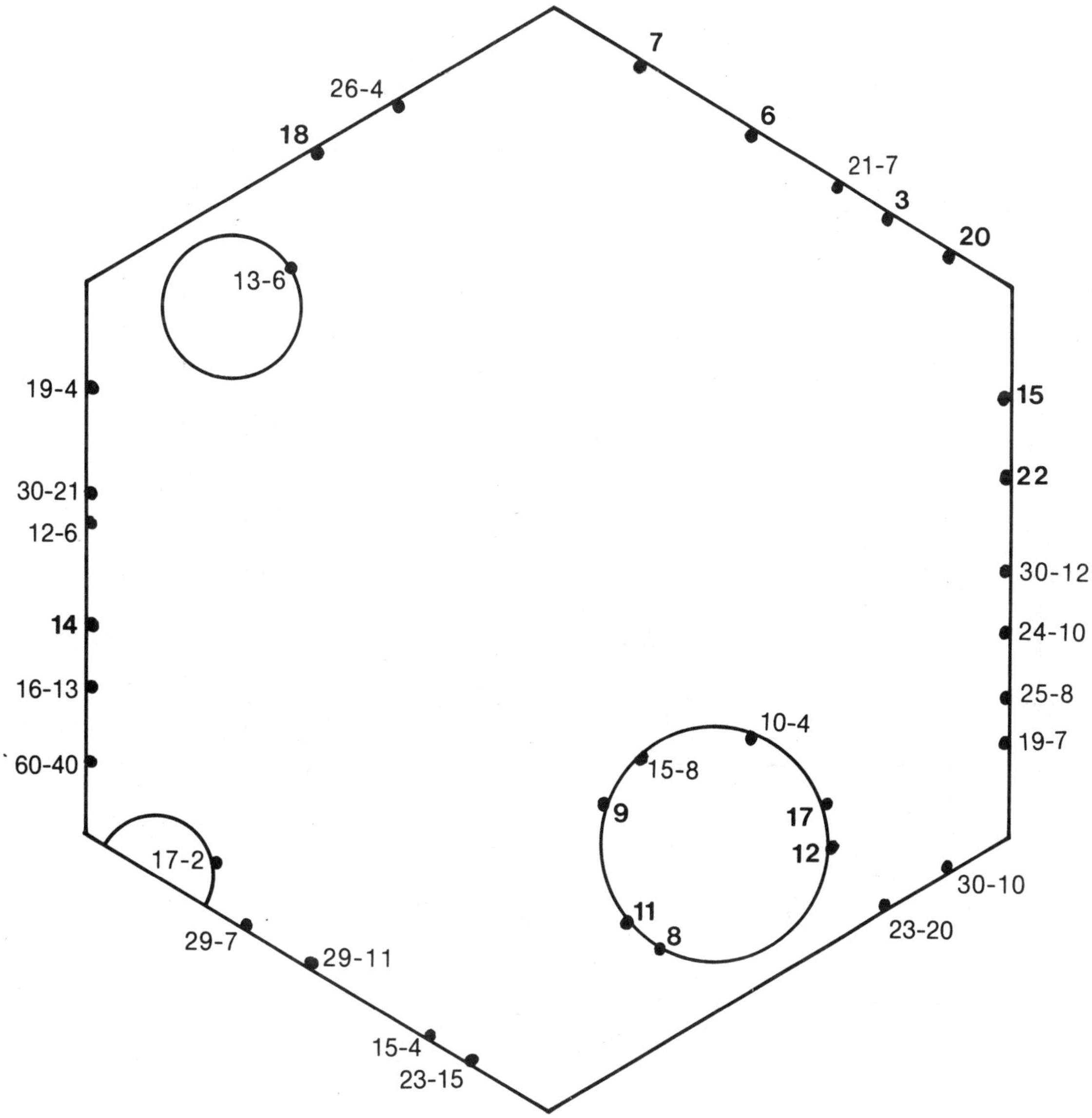

Correction Key
Design #14

1. Use this design as a correction key.
2. Allow your students to correct their own work.
3. Make a transparency of this design and instruct your students to place the transparency over their completed design for a quick and easy check.

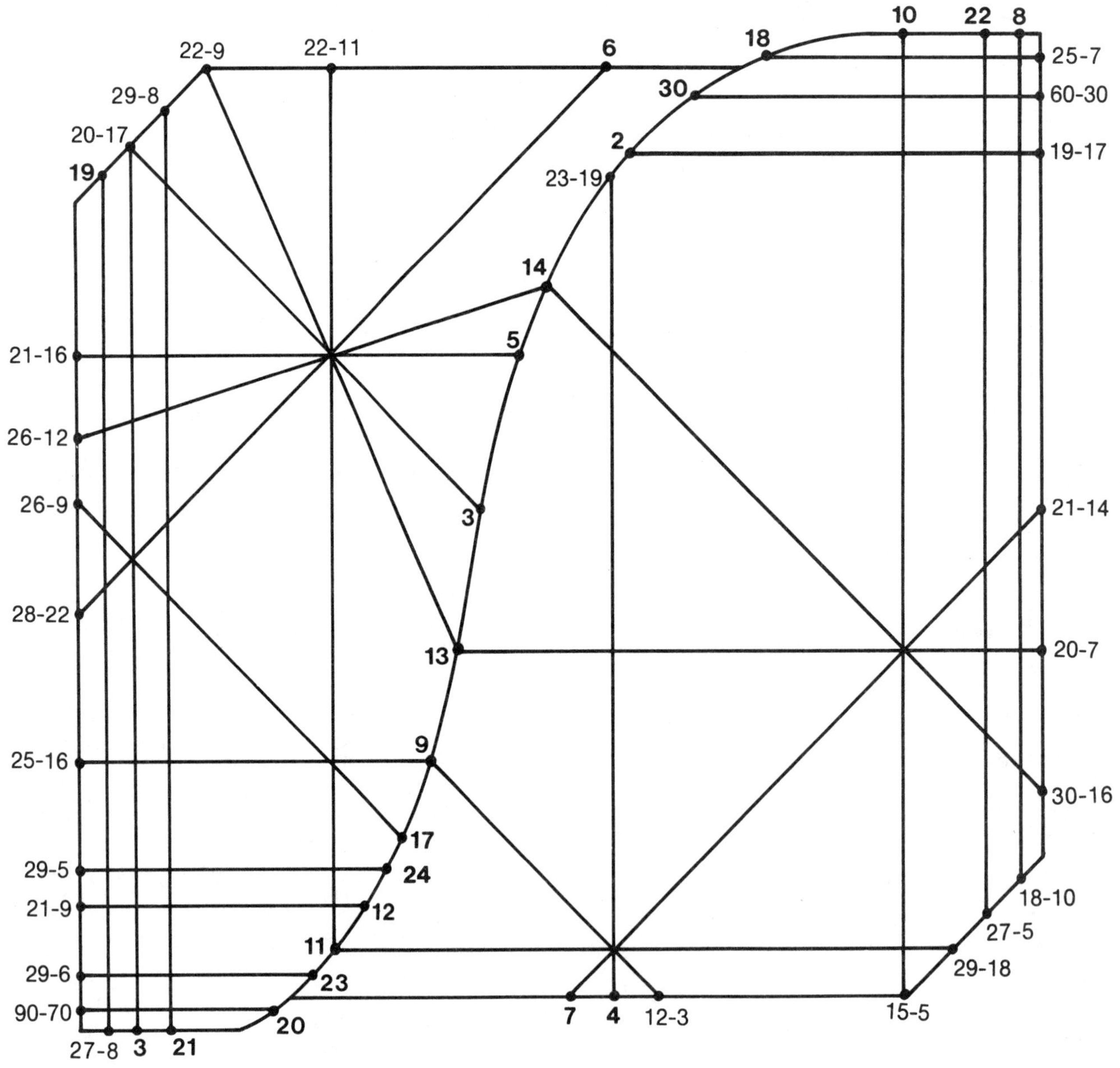

Subtraction
Design #14

Name ______________________

Date ______________________

1. **Read all of the directions before you begin.**
2. **Complete one subtraction problem next to a dot.**
   (You can use another piece of paper if you need to.)
3. **After you complete the problem, find the answer on the paper.**
4. **Use a ruler (or straight edge) and draw a line to connect the dot by the problem to the dot by the correct answer.**
5. **Complete each of the problems in the same way.**
6. **Make sure that you complete all of the problems. Some answers may be used more than once.**
7. **Carefully color your design after completing *all* of the problems.**

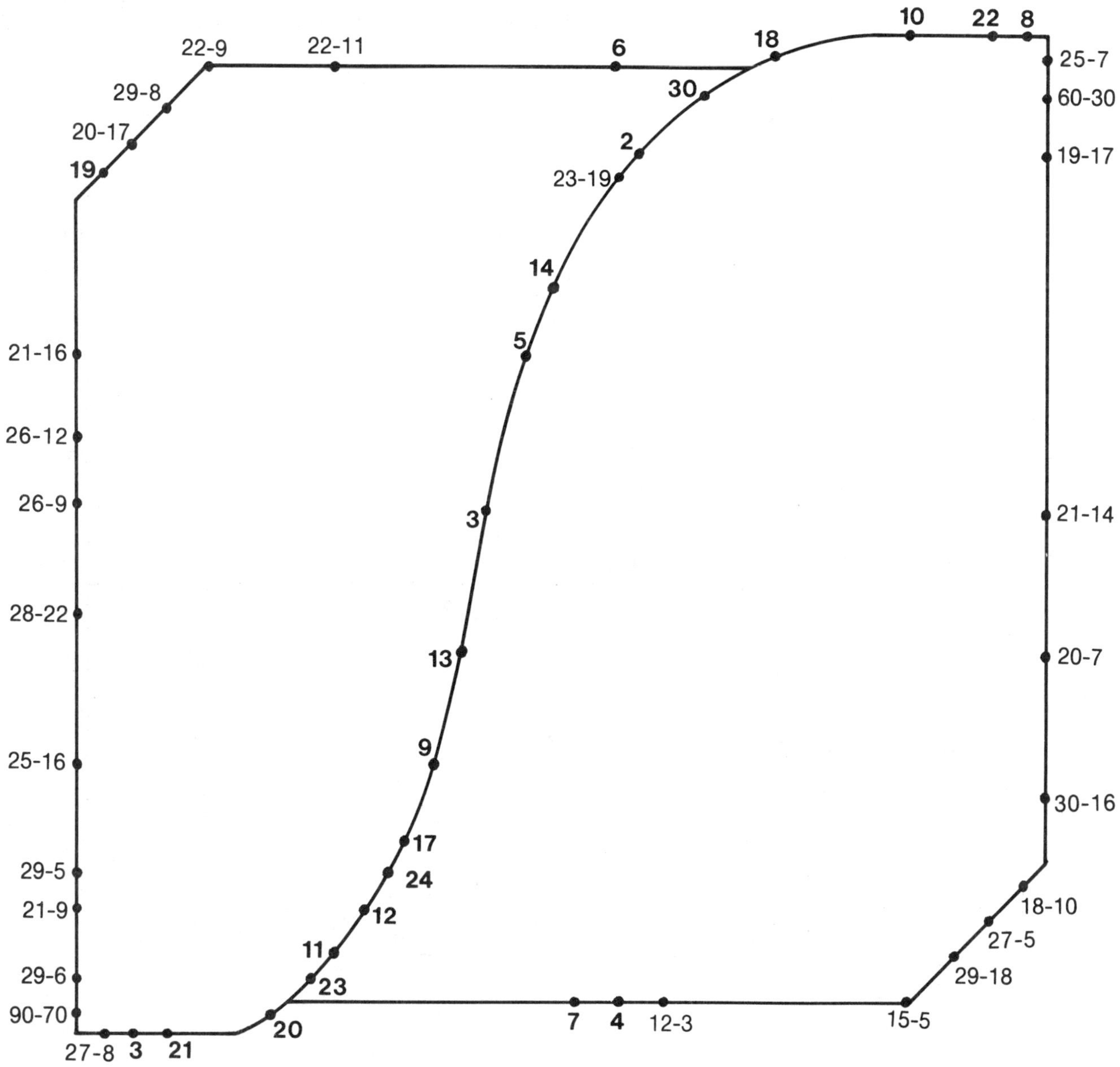

Correction Key
Design #15

1. Use this design as a correction key.
2. Allow your students to correct their own work.
3. Make a transparency of this design and instruct your students to place the transparency over their completed design for a quick and easy check.

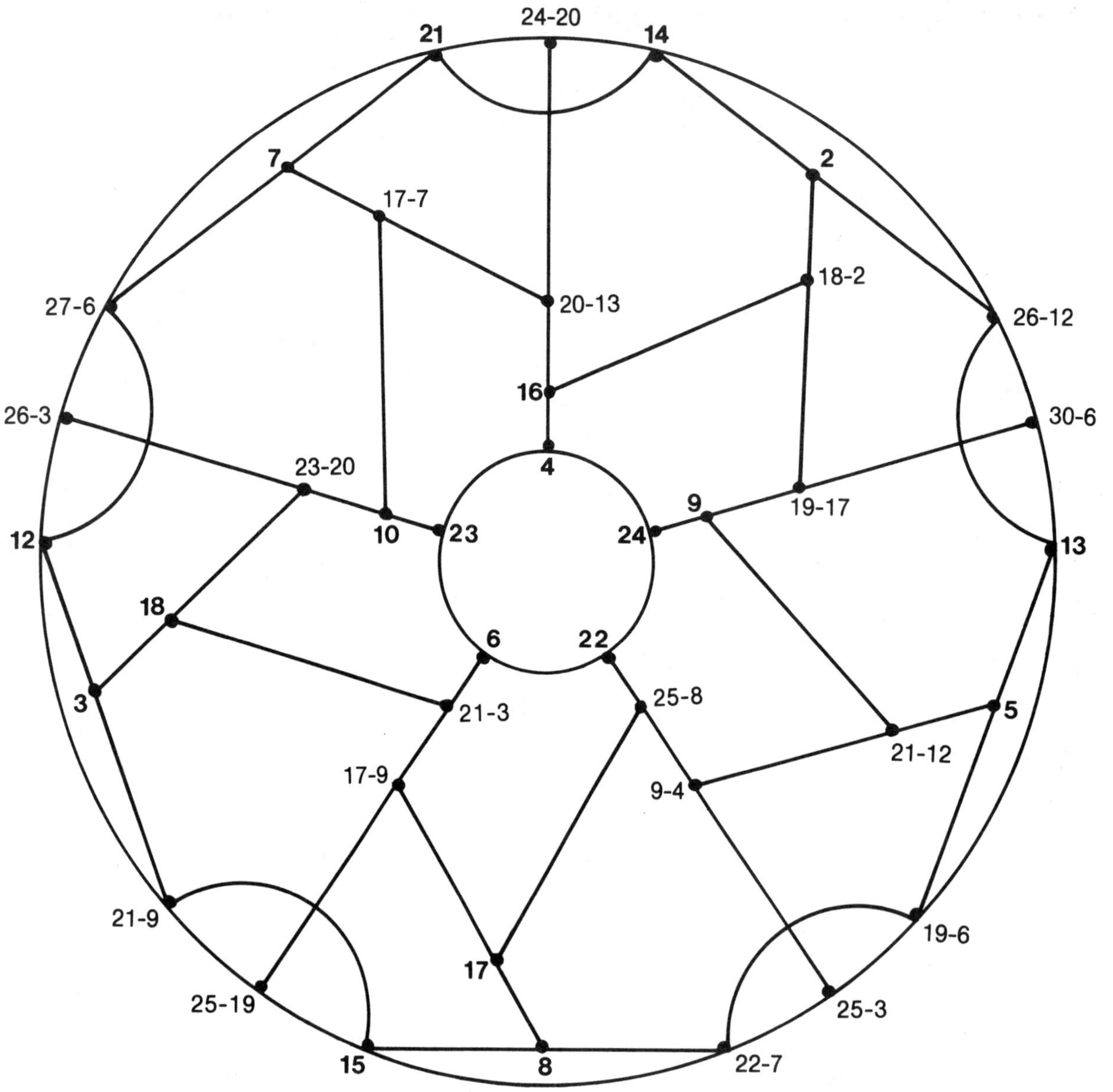

Subtraction
Design #15

Name ______________________

Date ______________________

1. Read all of the directions before you begin.
2. Complete one subtraction problem next to a dot.
   (You can use another piece of paper if you need to.)
3. After you complete the problem, find the answer on the paper.
4. Use a ruler (or straight edge) and draw a line to connect the dot by the problem to the dot by the correct answer.
5. Complete each of the problems in the same way.
6. Make sure that you complete all of the problems.
   Some lines pass through dots that are not part of the problem.
7. Carefully color your design after completing *all* of the problems.

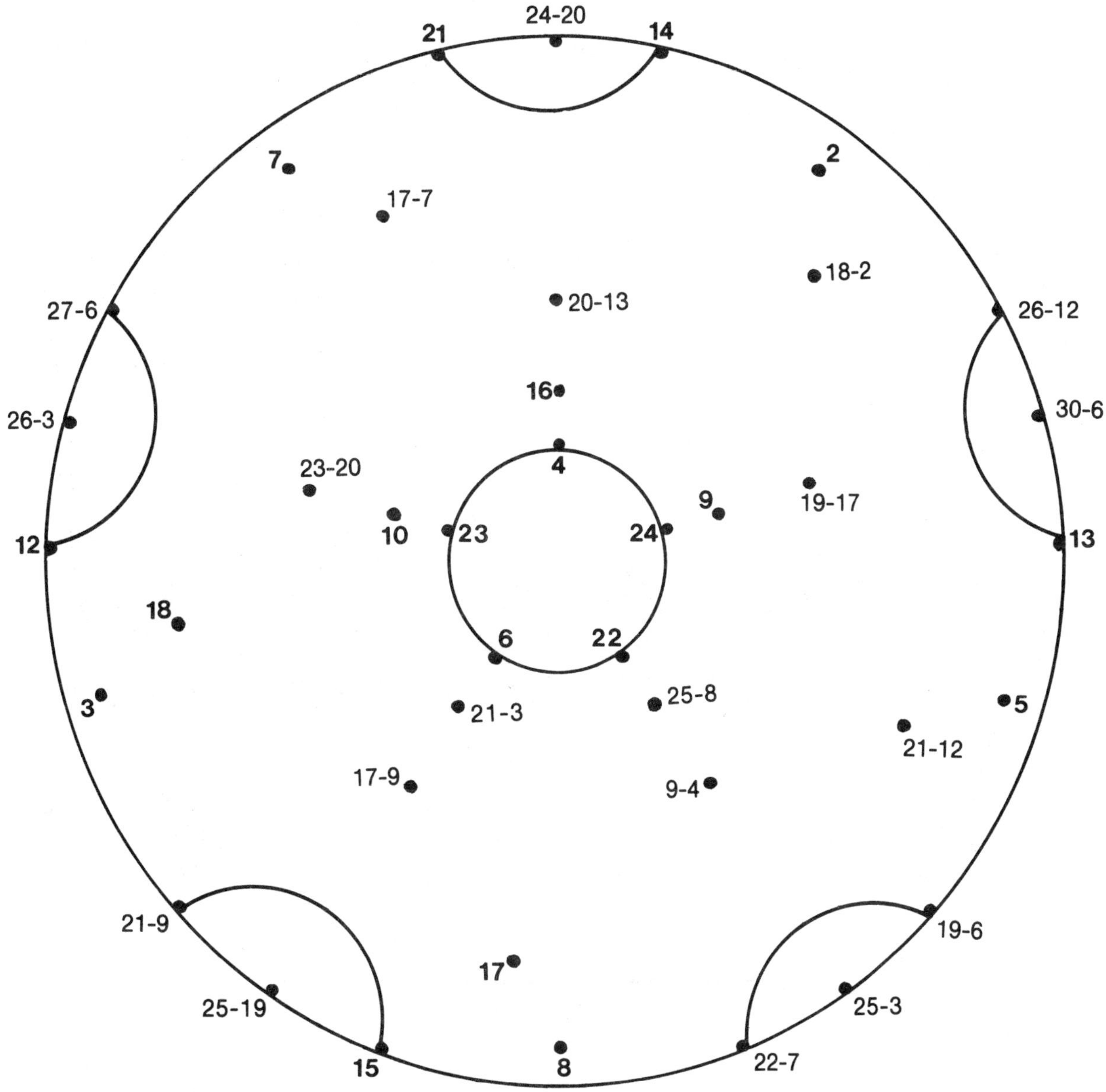

Correction Key
Design #16

1. Use this design as a correction key.
2. Allow your students to correct their own work.
3. Make a transparency of this design and instruct your students to place the transparency over their completed design for a quick and easy check.

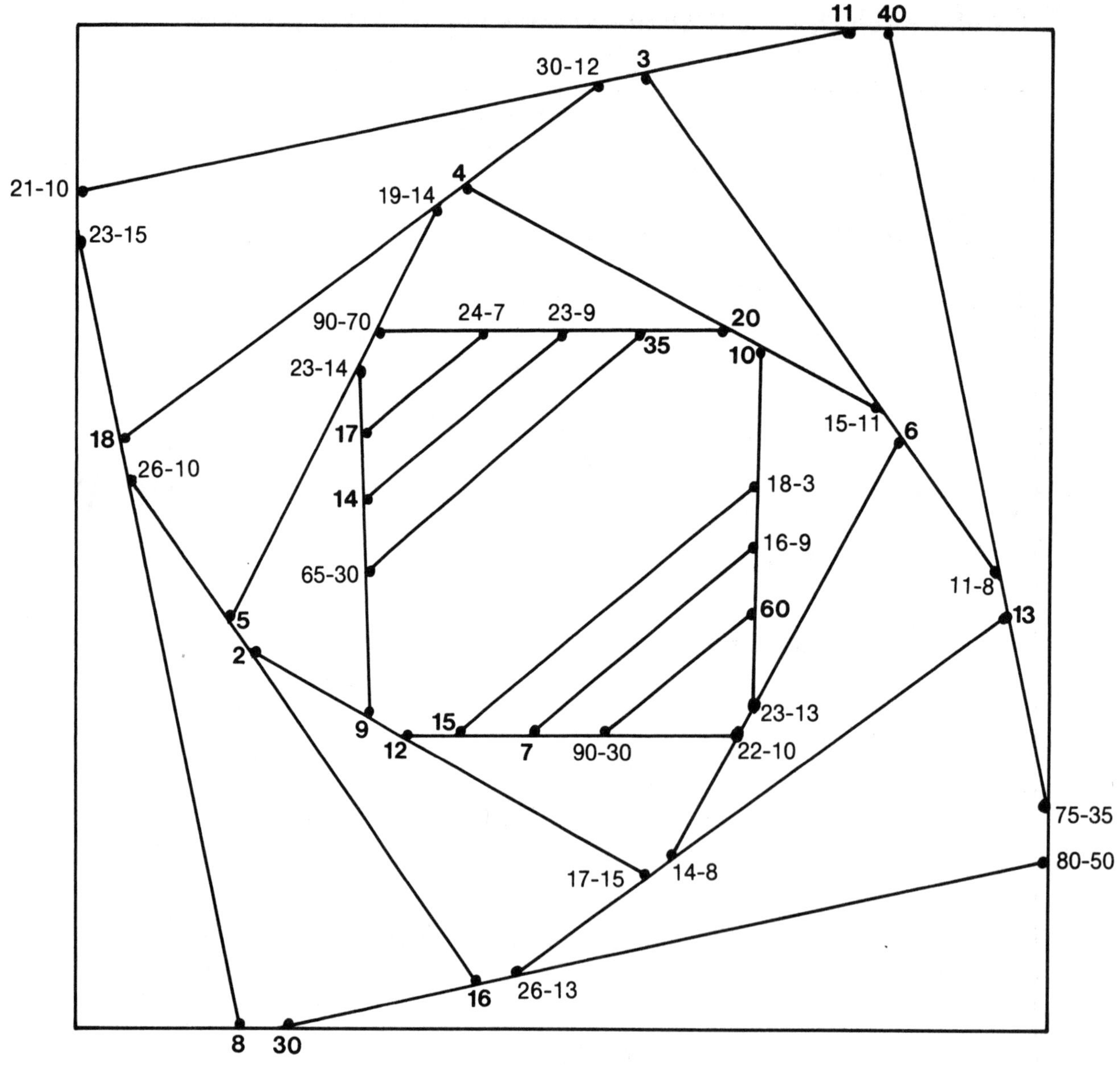

Subtraction
Design #16

Name ______________________________

Date ______________________________

1. Read all of the directions before you begin.
2. Complete one subtraction problem next to a dot. (You can use another piece of paper if you need to.)
3. After you complete the problem, find the answer on the paper.
4. Use a ruler (or straight edge) and draw a line to connect the dot by the problem to the dot by the correct answer.
5. Complete each of the problems in the same way.
6. Make sure that you complete all of the problems. Some lines pass through dots that are not part of the problem.
7. Carefully color your design after completing *all* of the problems.

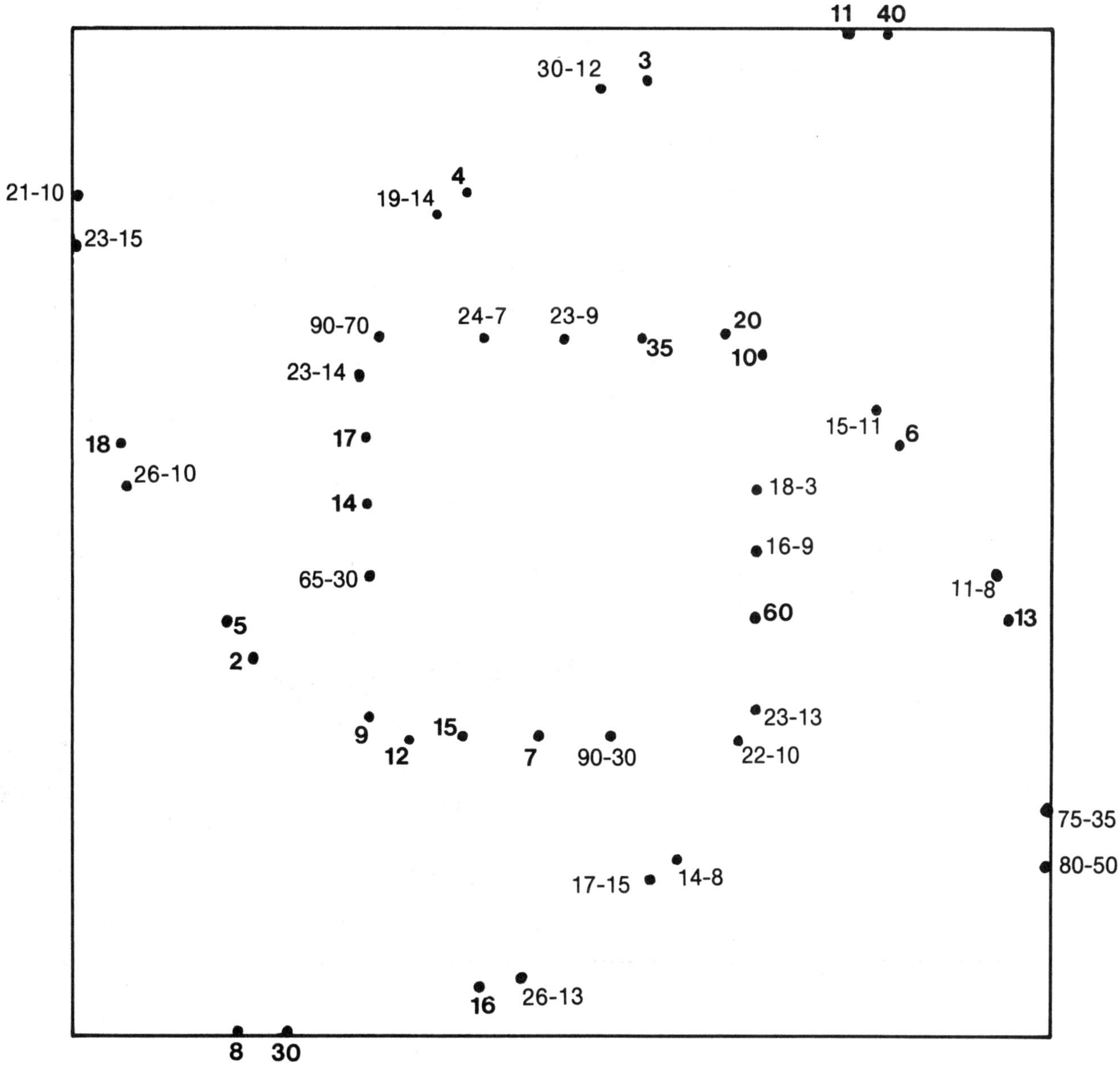

Correction Key
Design #17

1. Use this design as a correction key.
2. Allow your students to correct their own work.
3. Make a transparency of this design and instruct your students to place the transparency over their completed design for a quick and easy check.

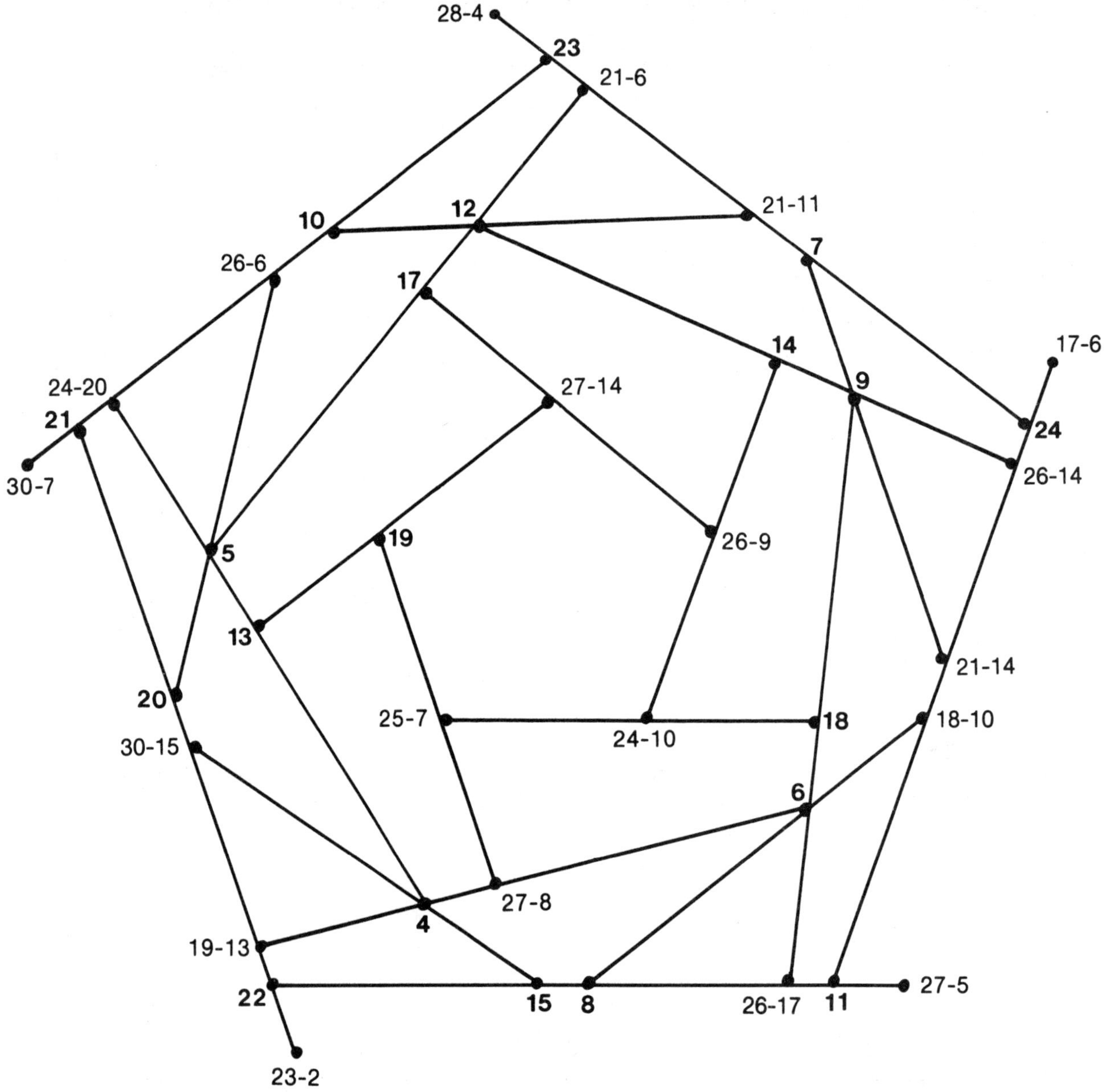

Subtraction
Design #17

Name ________________________________

Date ________________________________

1. Read all of the directions before you begin.
2. Complete one subtraction problem next to a dot.
   (You can use another piece of paper if you need to.)
3. After you complete the problem, find the answer on the paper.
4. Use a ruler (or straight edge) and draw a line to connect the dot by the problem to the dot by the correct answer.
5. Complete each of the problems in the same way.
6. Make sure that you complete all of the problems.
   Some lines pass through dots that are not part of the problem.
7. Carefully color your design after completing *all* of the problems.

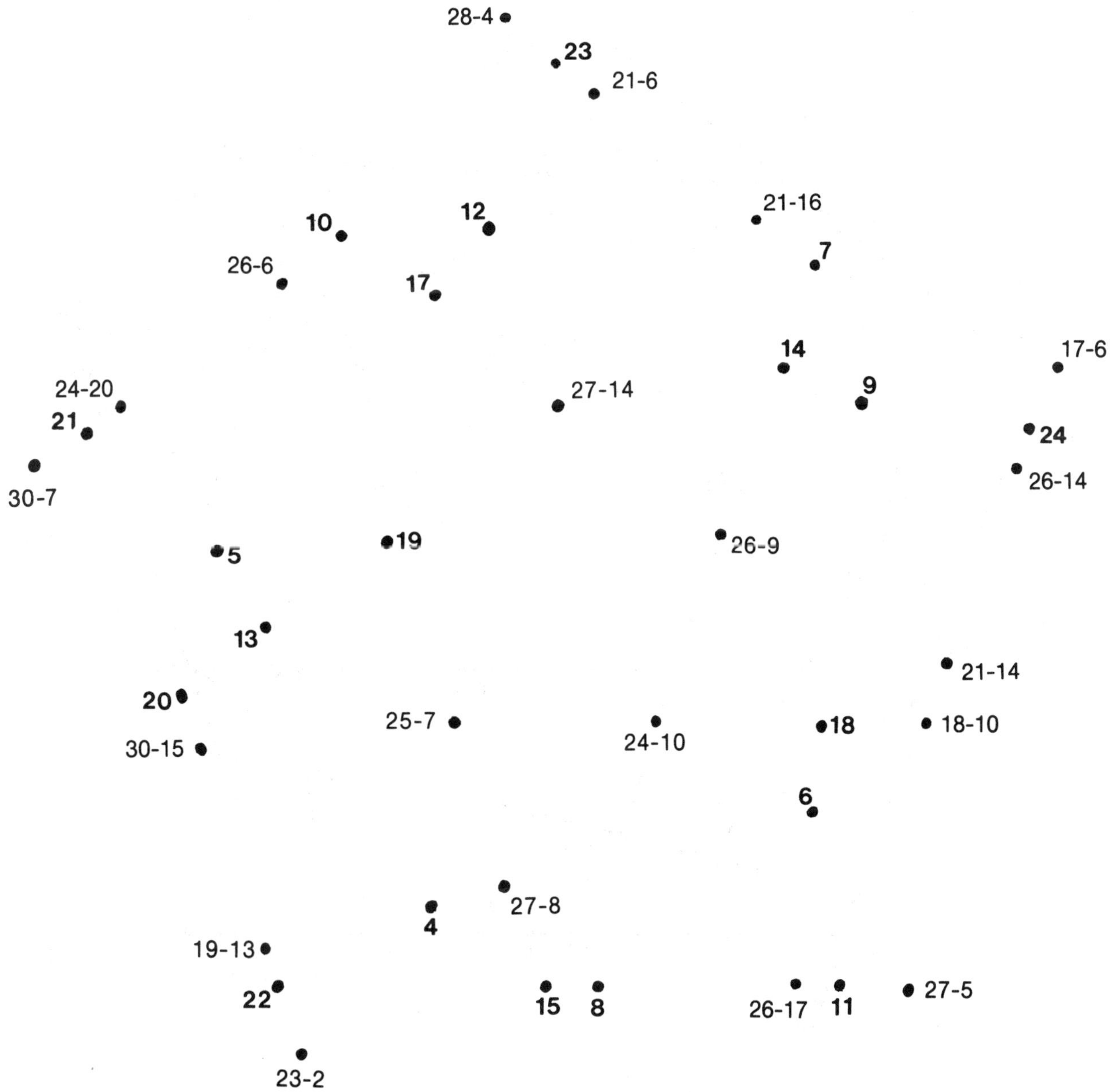

Correction Key
Design #18

1. Use this design as a correction key.
2. Allow your students to correct their own work.
3. Make a transparency of this design and instruct your students to place the transparency over their completed design for a quick and easy check.

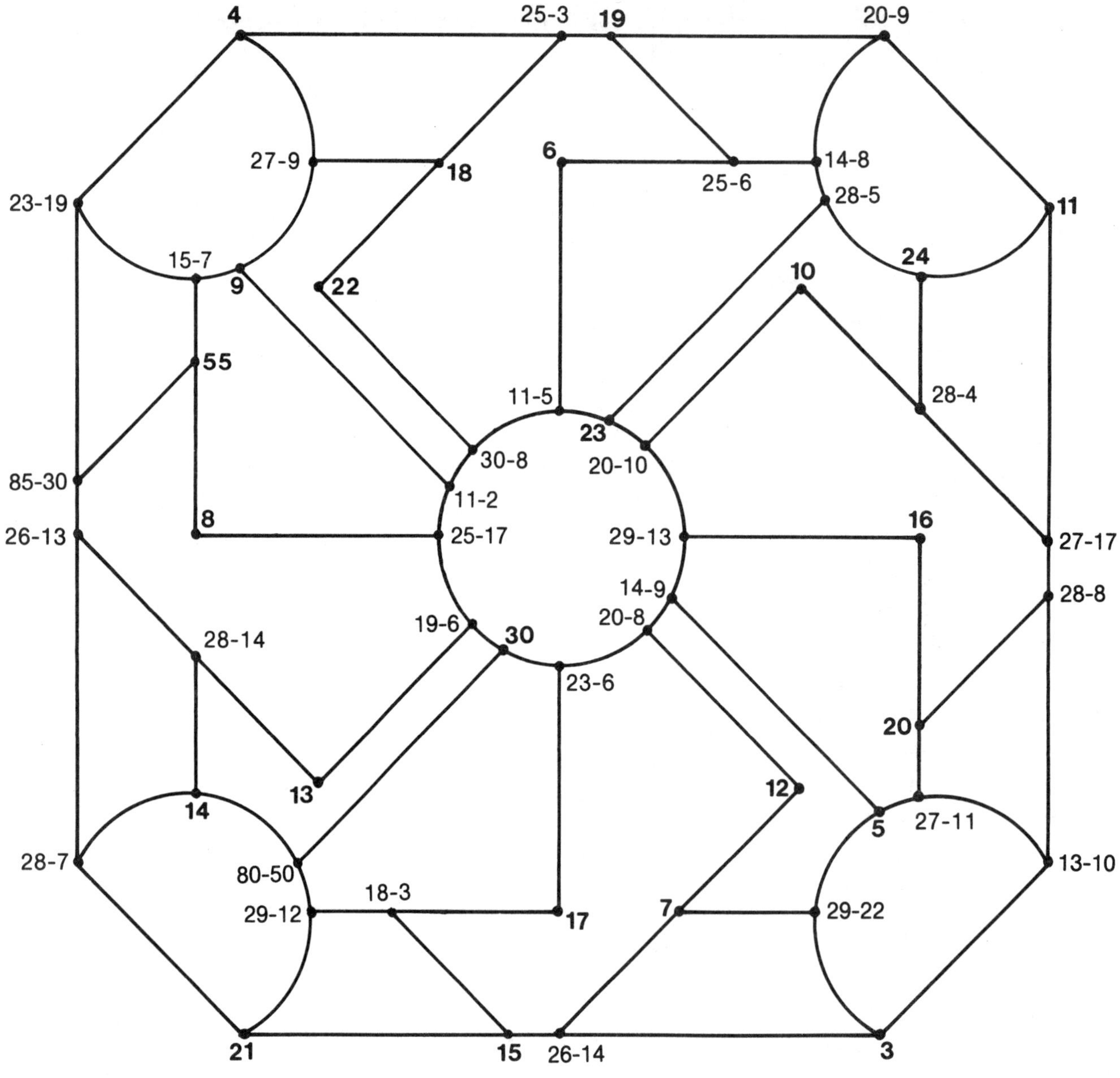

Subtraction

Design #18

Name ______________________________

Date ______________________________

1. **Read all of the directions before you begin.**
2. **Complete one subtraction problem next to a dot.**
   (You can use another piece of paper if you need to.)
3. **After you complete the problem, find the answer on the paper.**
4. **Use a ruler (or straight edge) and draw a line to connect the dot by the problem to the dot by the correct answer.**
5. **Complete each of the problems in the same way.**
6. ✏ **Be careful — some lines pass through dots that are not part of the problem. Also, some answers may be used more than once.**
7. **Carefully color your design after completing *all* of the problems.**

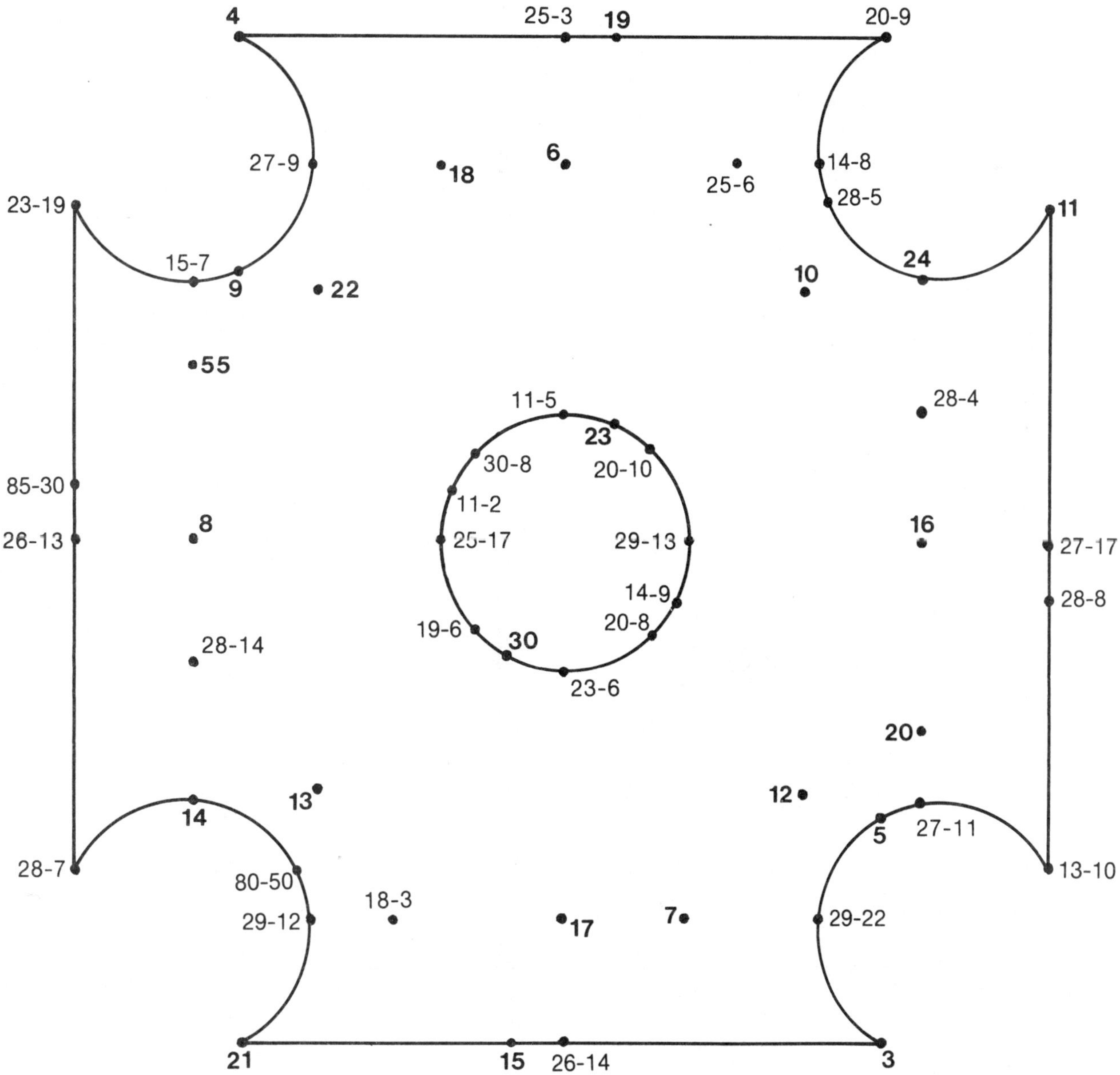

Correction Key
Design #19

1. Use this design as a correction key.
2. Allow your students to correct their own work.
3. Make a transparency of this design and instruct your students to place the transparency over their completed design for a quick and easy check.

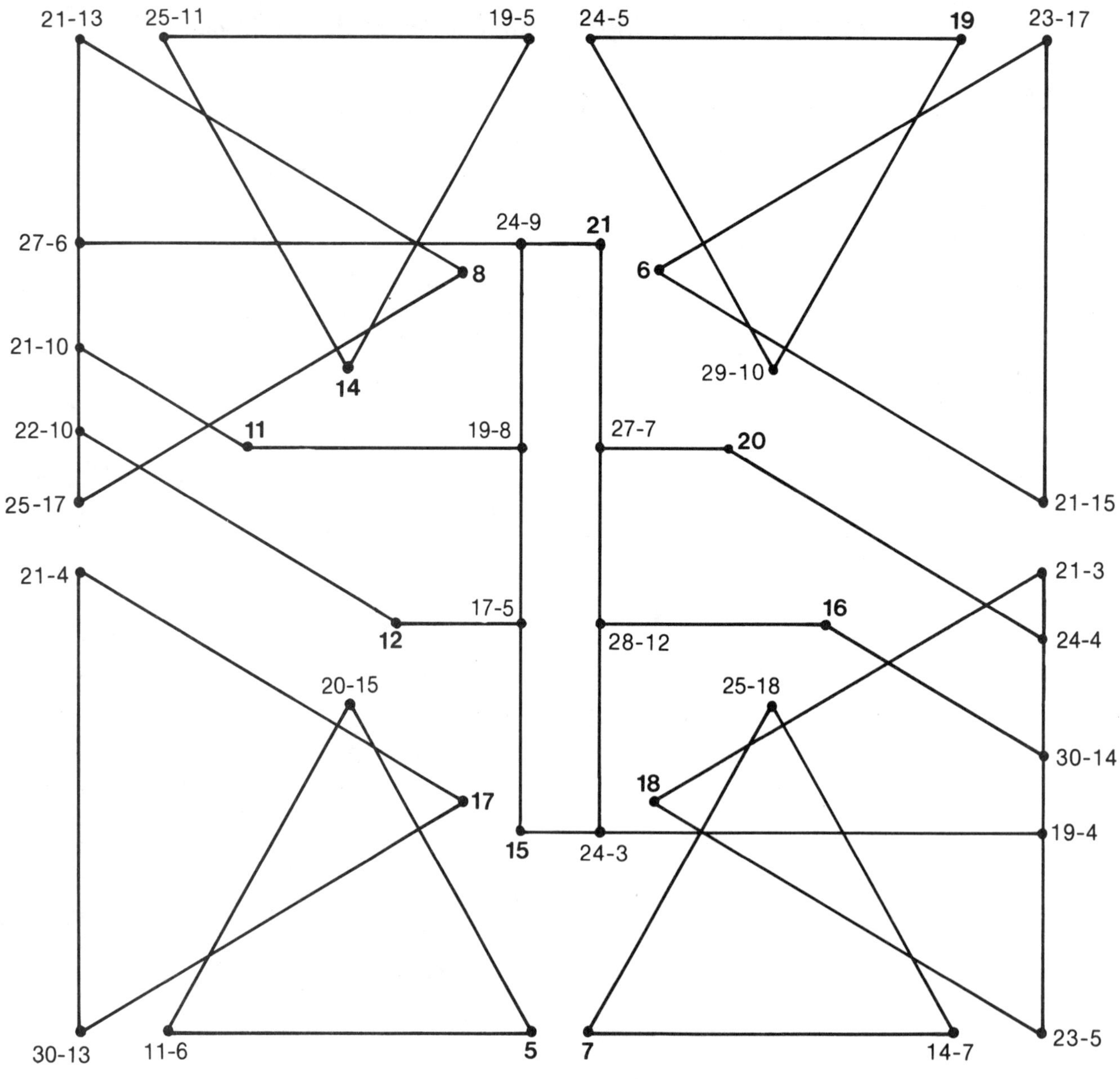

Subtraction
Design #19

Name ______________________

Date ______________________

1. **Read all of the directions before you begin.**
2. **Complete one subtraction problem next to a dot.**
   (You can use another piece of paper if you need to.)
3. **After you complete the problem, find the answer on the paper.**
4. **Use a ruler (or straight edge) and draw a line to connect the dot by the problem to the dot by the correct answer.**
5. **Complete each of the problems in the same way.**
6. ✏ **Be careful — some lines pass through dots that are not part of the problem. Also, some answers may be used more than once.**
7. **Carefully color your design after completing *all* of the problems.**

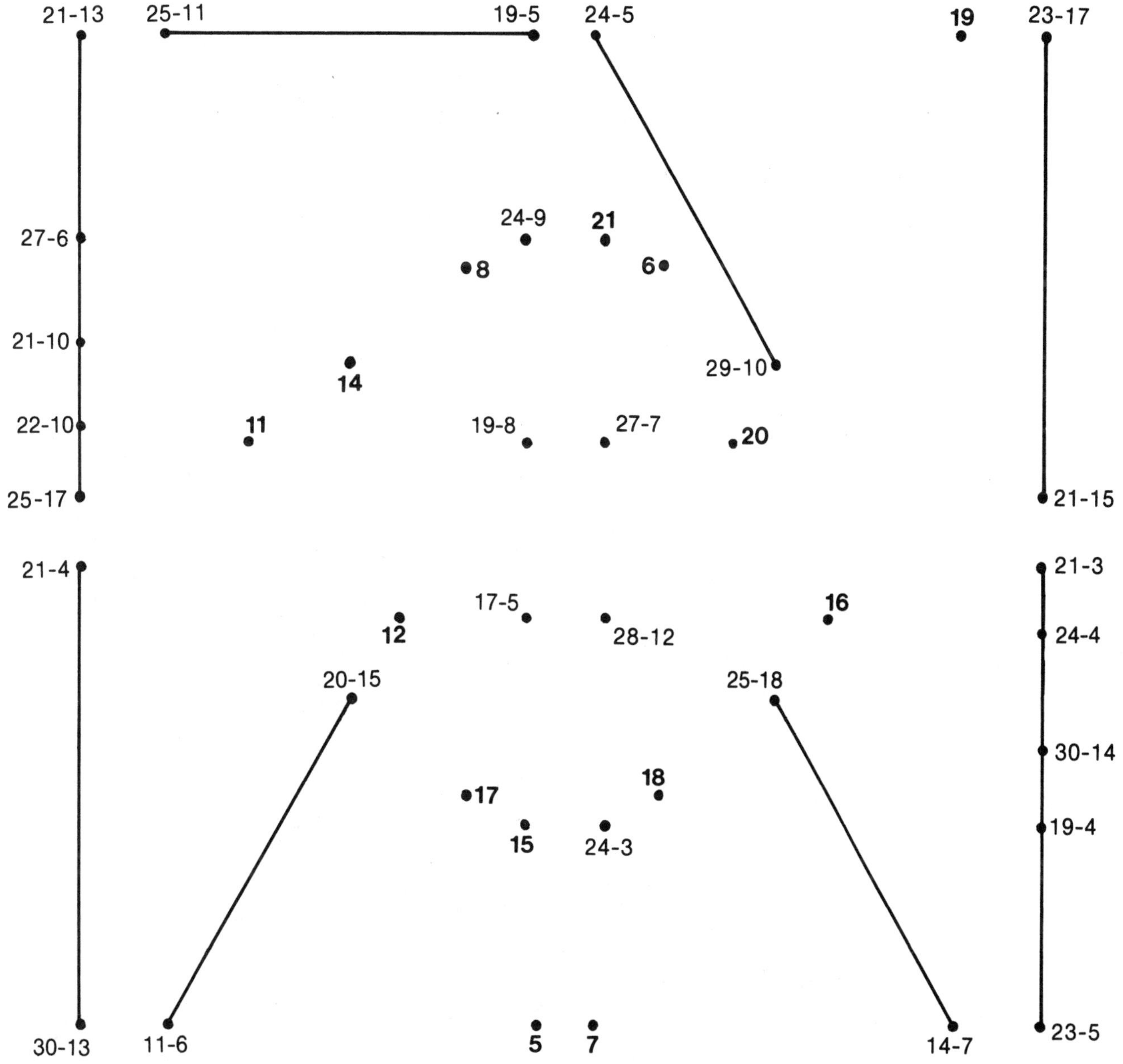

Correction Key
Design #20

1. Use this design as a correction key.
2. Allow your students to correct their own work.
3. Make a transparency of this design and instruct your students to place the transparency over their completed design for a quick and easy check.

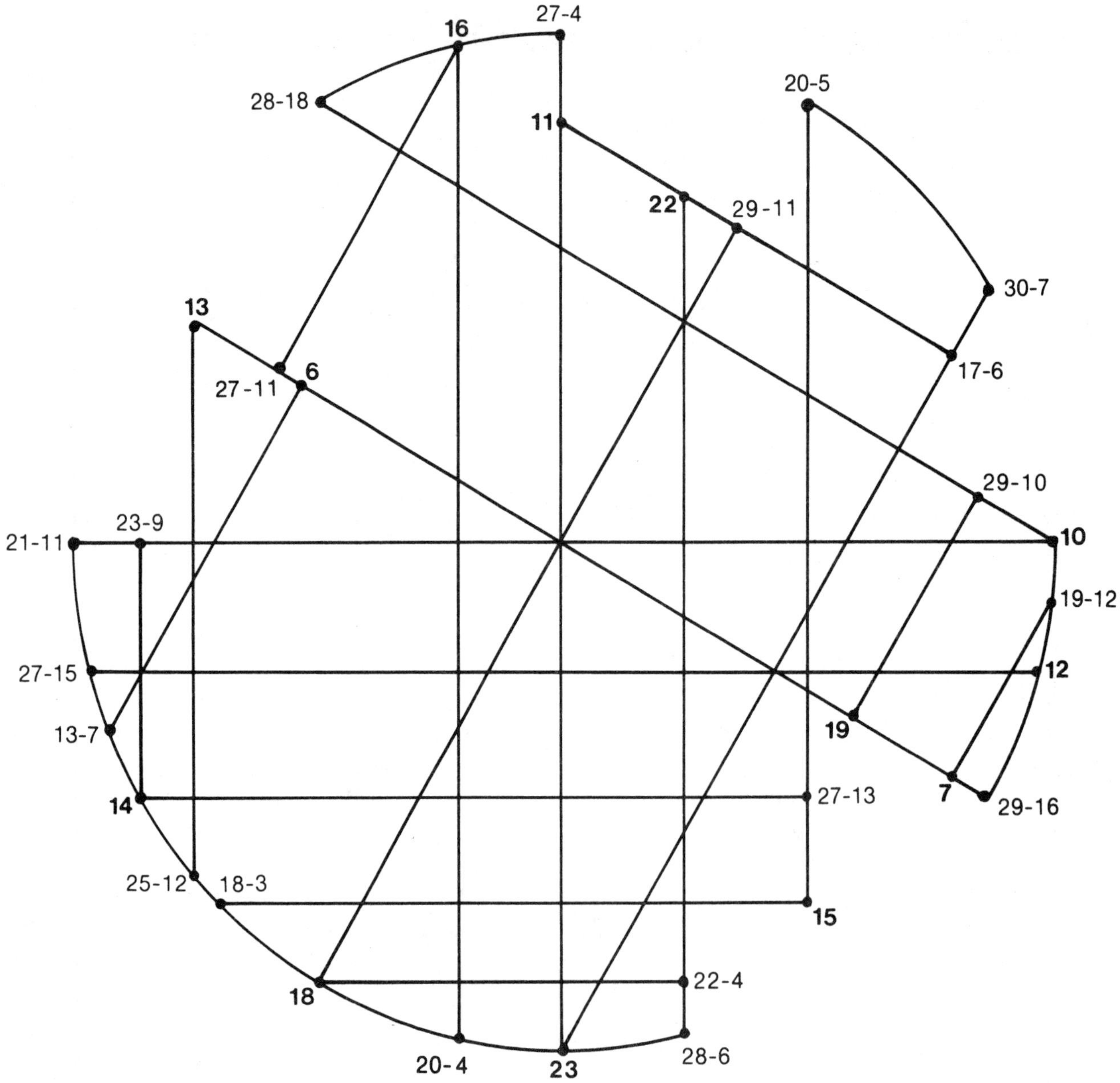

Subtraction
Design #20

Name ______________________________

Date ______________________________

1. **Read all of the directions before you begin.**
2. **Complete one subtraction problem next to a dot.**
   (You can use another piece of paper if you need to.)
3. **After you complete the problem, find the answer on the paper.**
4. **Use a ruler (or straight edge) and draw a line to connect the dot by the problem to the dot by the correct answer.**
5. **Complete each of the problems in the same way.**
6. ✏ **Be careful — some lines pass through dots that are not part of the problem. Also, some answers may be used more than once.**
7. **Carefully color your design after completing *all* of the problems.**

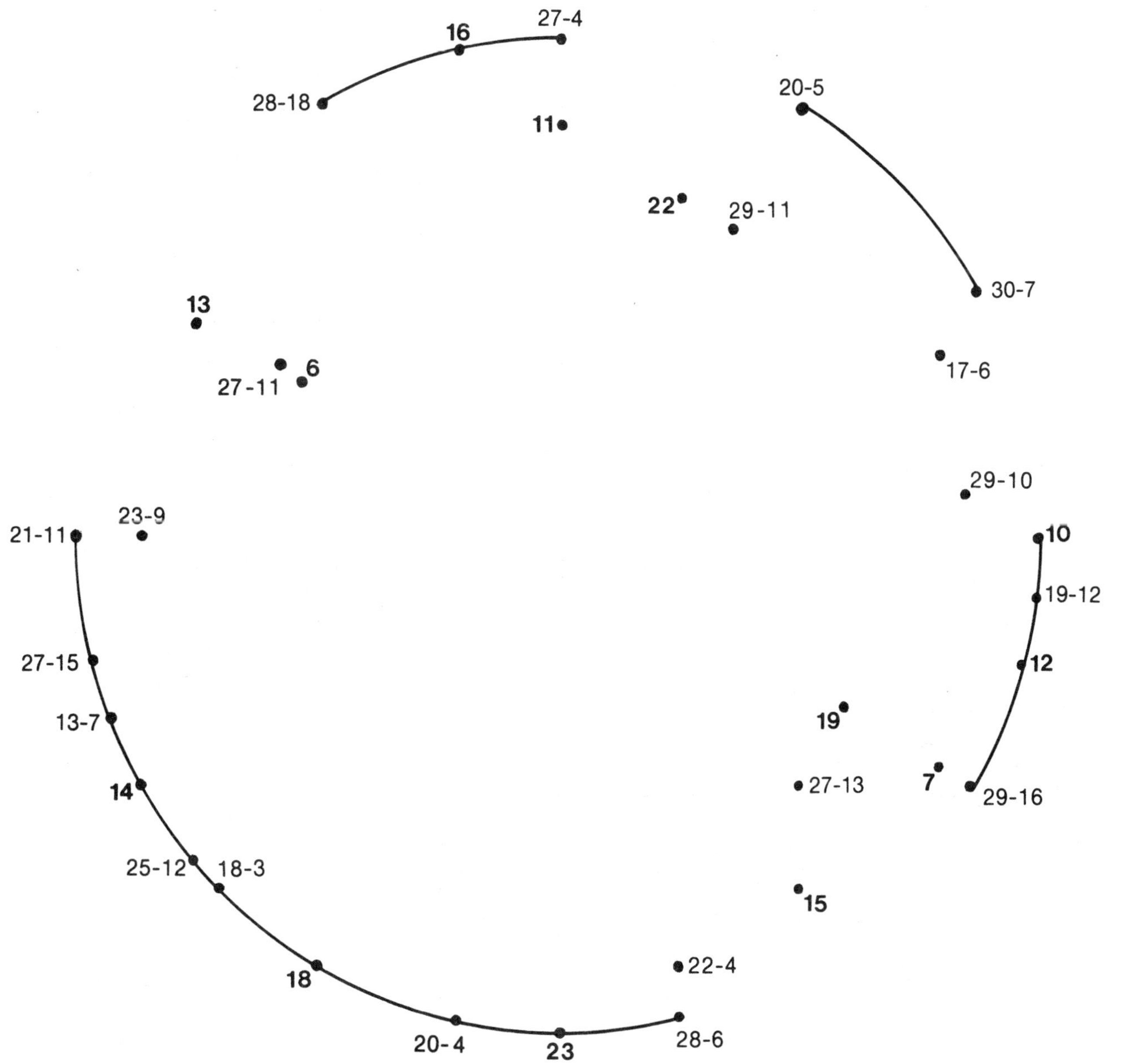

Creating Line Designs

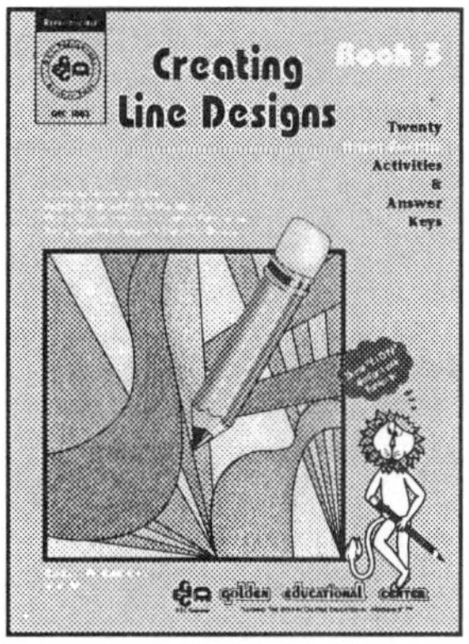

Best Seller

Students use a straight edge to sequentially connect dots in order to create geometric designs. These activities develop visual perception and eye-hand coordination as they build students' self-confidence. Each book has twenty activities and completed designs, which can be used for the answer key. *Teacher instructions are included.*

Book 1 #1001 . . . . K-1st
Book 2 #1002 . . . . 1st-3rd
Book 3 #1003 . . . . 3rd-5th
Book 4 #1004 . . . . 4th-7th

Designs in Math

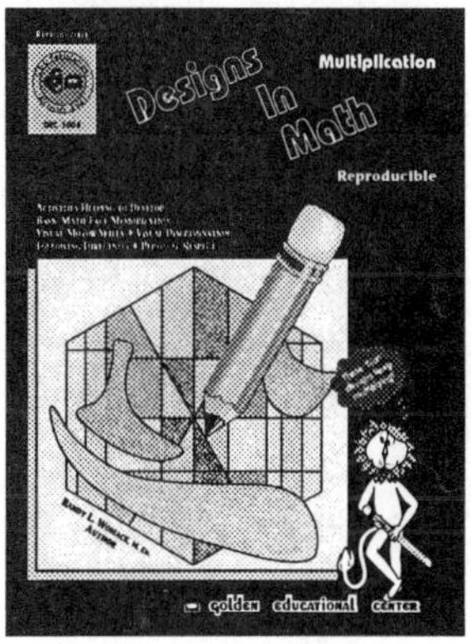

Best Seller

Students create geometric designs by connecting the dots between math problems and the correct answers. These activities reinforce basic math fact memorization, following directions and principles of design. Each book contains 20 activity pages with one design per page. There is also a completed design that can be used for a correction key or made into a transparency for classroom instruction. *Teacher instructions are included.*

Addition #1006 . . . . 3rd-5th
Subtraction #1007 . . . . 3rd-5th
Multiplication #1008 . . . . 4th-6th
Division #1009 . . . . 4th-7th
Fractions #1010 . . . . 5th-9th
Frac.-Dec. Equivalents #1011 5th-9th

❖ ❖ ❖

Beginning Math Art

Best Seller

Similar to *Designs in Math,* these books show students the fundamentals of design while they practice solving their math problems and then connect the correct dots with a straight edge. There is a maximum of 12 problems per design. Young students will enjoy completing the 20 activity pages in each of these books. *Teacher instructions and answer keys are included.*

Add & Subtract 0-10 #1013 K-1st
Add & Subtract 11-20 #1014 1st-2nd
Multiply & Divide 0-12 #1015 2nd-3rd

Read✦N✦Draw
Following Directions

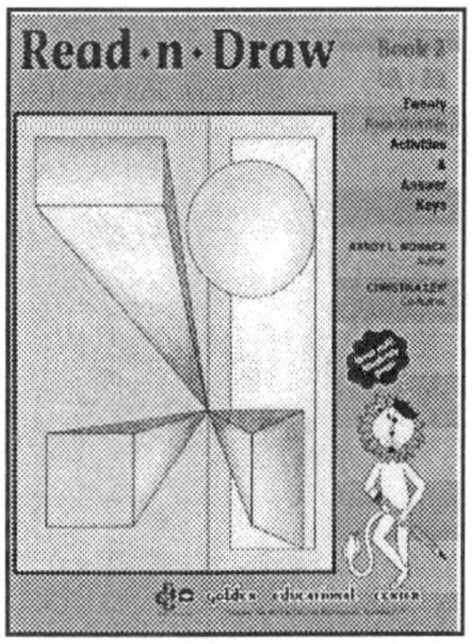

Best Seller

These workbooks teach children to follow directions through exciting reading and measuring activities. Students build reading comprehension, as well as measuring skills, by reading directions, plotting and then connecting points with a ruler to create interesting designs. Twenty activities in each book. *Teacher instructions, students' glossary of new terms & keys are included.*

Book 1 #1021 . . . . 3rd-5th
Book 2 #1022 . . . . 4th-8th

❖ ❖ ❖

Using a Visual Grid for Solving
Math Word Problems

New

Paring down to the most essential information, young students can now see and even understand the most essential elements/terms of a word problem. There are 35 lessons in each of these books. *Teacher instructions and answer keys are included.*

Add & Subtract 0-99 #1112
(no borrowing/carrying) K-2nd
Add & Subtract to 122 #1113
(with borrowing/carrying) 1st-3rd

❖ ❖ ❖ ❖ ❖ ❖ ❖ ❖ ❖ ❖ ❖ ❖ ❖ ❖ ❖

Other books available from Golden Educational Center. All of these books have an easy-to-read & use format.
Contact your local retail outlet or Golden Educational Center .

N. America Country Studies #1965 . 4th & Above
S. America Country Studies #1975 . 4th & Above
Far East Country Studies #1935 . . . 4th & Above
Mid-East Country Studies #1936 . . . 4th & Above
Canada Province Studies #1985. . . . 4th & Above
Calif. Early History #2911. . . . . 4th & Above
Calif. Geography #2912 . . . . 4th & Above
Wash. State Geography #2106 . . . . 4th & Above
U.S. Outline Maps & Studies #1992 4th & Above
U.S. Geography #1993 . . . . 4th & Above
New **Complete U.S. State Studies** #1995 . 4th & Above
Continent Maps & Studies #1905. . . 4th & Above
Learning the Continents #1906. . . . 4th & Above
Ecology Activities #1205 . . . . K-3rd
Ecology Math #1105 . . . . 5th-8th
Creating Calendars #1201 . . . . Pre K-3rd
Creating Holiday #1202 . . . . Pre K-3rd

Learning the Continents

Students use maps to identify, memorize and locate the countries, waterways and points of interest on each continent (except Antarctica). Each continent has 16 pages of activities, including a word search for review. *Teacher instructions and keys are included.*

#1906 .... 4th-8th

Best Sellers

Country Study Books

Each individual country within each respective continent has new words to look up in the dictionary, a large outline map, one page of current facts (population, area, capital, largest city, flag w/description, additional interesting information, and more), one or two pages of history and a page of review questions at the end of each country's section. *Teacher instructions and answer keys are included.*

No. America #1965 .... 4th-8th

So. America #1975......... 4th-8th

Far East #1935 .... 4th-8th

Middle East #1936 .... 4th-8th

Canada #1985 .... 4th-8th

Continent Maps & Studies

This book contains Outline, Waterway, Political Boundary Map and Individual Fact Sheet for all of the continents. (Antarctica only has fact sheet and outline map.) There are questions, research activities and a glossary that can be used with each continent. *Teacher instructions and keys are included.*

#1905.................. 4th-8th

❖ ❖ ❖

California Early History: This is a simplified, yet complete, resource book detailing the history of California through statehood. This tremendous resource includes sections on the contributions made by Native Ameri-cans, Asians, Africans and Mexicans to the growth and development of the state. Review questions and bonus activities follow each section. *94 pages; Teacher instructions and keys are included.*

California Geography: Students are given a world overview with a specific review of California's climate and its physical, economic and political features. Lessons are reinforced with maps, exercises, review questions and bonus activities.
63 pages; Teacher instructions and answer keys are included.

Washington Geography: Students are given a world overview with a specific review of Washington's climate and its physical, economic and political features. Lessons are reinforced with maps, exercises, review questions and bonus activities.
63 pages; Teacher instructions and answer keys are included.

California Early History #2911 .... 4th-8th

California Geography #2911 .... 4th-8th

Washington Geography #2102 .... 4th-8th

❖ ❖ ❖

U.S. Studies

Best Sellers

U.S. Outline Maps: This book has an individual fact sheet and outline map for each of the 50 states, Washington, D.C. and the entire United States. U.S. waterways and state boundary maps are also included. There are also individual question and research pages that can be used with each of the states.
110 pages with teacher instructions

U.S. Geography: Each section of this book begins with new vocabulary words to look up and define. The sections simply cover a world's overview, physical, economic, political and climatic features of the U.S. Maps, activities, questions and review questions are included in each section.
63 pages; Teacher instructions and answer keys are included.

Complete Book of U.S. State Studies: **NEW**
This book includes the same sections as in the *U.S. Outline Maps* book listed above. It also includes historical information about the Native Americans living in the region before any settlers came, as well as the early settlers up to the time of statehood. Each state also has interesting trivia facts and a word search puzzle.
219 pages; Teacher instructions and answer keys are included.

U.S. Outline Maps #1992 .... 4th-8th

U.S. Geography #1993 .... 4th-8th

Complete U.S. State Studies #1995.................... 4th-8th